Progress in the Chemistry of Organic Natural Products

The volumes of this classic series, now referred to simply as "Zechmeister" after its founder, Laszlo Zechmeister, have appeared under the Springer Imprint ever since the series' inauguration in 1938. It is therefore not really surprising to find out that the list of contributing authors, who were awarded a Nobel Prize, is quite long: Kurt Alder, Derek H. R. Barton, George Wells Beadle, Dorothy Crowfoot-Hodgkin, Otto Diels, Hans von Euler-Chelpin, Paul Karrer, Luis Federico Leloir, Linus Pauling, Vladimir Prelog, with Walter Norman Haworth and Adolf F. J. Butenandt serving as members of the editorial board.

The volumes contain contributions on various topics related to the origin, distribution, chemistry, synthesis, biochemistry, function or use of various classes of naturally-occurring substances ranging from small molecules to biopolymers.

Each contribution is written by a recognized authority in the field and provides a comprehensive and up-to-date review of the topic in question. Addressed to biologists, technologists, and chemists alike, the series can be used by the expert as a source of information and literature citations and by the non-expert as a means of orientation in a rapidly developing discipline.

All contributions are listed in PubMed.

A. Douglas Kinghorn · Heinz Falk ·
Simon Gibbons · Yoshinori Asakawa · Ji-Kai Liu ·
Verena M. Dirsch
Editors

Tigliane Diterpenoids

Volume 125

By Giovanni Appendino and Simone Gaeta

Springer

Editors
A. Douglas Kinghorn
College of Pharmacy
Ohio State University
Columbus, OH, USA

Heinz Falk
Institute of Organic Chemistry
Johannes Kepler University Linz
Linz, Austria

Simon Gibbons
Natural and Medical Sciences Research Center
University of Nizwa
Nizwa, Oman

Yoshinori Asakawa
Faculty of Pharmaceutical Sciences
Tokushima Bunri University
Tokushima, Japan

Ji-Kai Liu
School of Pharmaceutical Sciences
South Central University for Nationalities
Wuhan, China

Verena M. Dirsch
Department of Pharmaceutical Sciences
University of Vienna
Vienna, Austria

ISSN 2191-7043 ISSN 2192-4309 (electronic)
Progress in the Chemistry of Organic Natural Products
ISBN 978-3-031-67179-1 ISBN 978-3-031-67180-7 (eBook)
https://doi.org/10.1007/978-3-031-67180-7

This Springer imprint is published by the registered company Springer Nature Switzerland AG
The registered company address is: Gewerbestrasse 11, 6330 Cham, Switzerland

About This Book

This volume deals with the distribution, chemistry, and molecular bioactivity of tiglianes, which are reviewed from the very beginning of the studies on these diterpenoids, summarizing their clinical and toxicological literature mostly in its more recent and controversial aspects, and critically analyzing various proposals for their biosynthesis.

Contents

Contributors

Giovanni Appendino Department of Pharmaceutical Sciences, Università del Piemonte Orientale, Novara, Italy

Simone Gaeta Research & Development—Chemistry Research, QBiotics Group Limited, Taringa, QLD, Australia

Tigliane Diterpenoids

Giovanni Appendino and Simone Gaeta

Contents

G. Appendino (✉)
Department of Pharmaceutical Sciences, Università del Piemonte Orientale, Largo Donegani, 2, 28100 Novara, Italy
e-mail: giovanni.appendino@uniupo.it

S. Gaeta
Research & Development—Chemistry Research, QBiotics Group Limited, 165, Moggill Road, Taringa, QLD 4068, Australia
e-mail: simone.gaeta@qbiotics.com

A. D. Kinghorn, H. Falk, S. Gibbons, Y. Asakawa, J.-K. Liu, V. M. Dirsch (eds.),
Tigliane Diterpenoids, Progress in the Chemistry of Organic Natural Products 125,
https://doi.org/10.1007/978-3-031-67180-7_1

1 Introduction

Phorbol! Woodward started his 1963 article on the total synthesis of strychnine with the veritable equivalent of the four-note (short-short-short-long) motif of Beethoven's Fifth Symphony ("Strychnine!") [1]. Since the skin irritancy and blistering properties of phorbol esters are in no way less legendary than the toxicity of strychnine, a similar emphatic entry seems suitable for a review on tiglianes, the class of diterpenoids to which phorbol (**1**) belongs. This polyol is one of the most well-documented natural products in the whole biomedical literature, with a search on PubMed totaling 53,188 entries as of March 1, 2024. For comparison, Taxol (paclitaxel) had 48,391 entries, artemisinin 11,822, strychnine 6,756, and cannabidiol 6,525, to name just a few popular plant natural products. Despite all this attention, phorbol (**1**) is not at all a literature "commodity", and the way it is produced by plants and acts in cells and living organisms is still far from clear, while a bird's eye view on the phorbol/humans interface reveals a puzzling overlapping of poisonous, medicinal, and cosmetic uses for phorbol myristate acetate (PMA) (**2**), its archetypal ester. This compound has also been referred to in the literature as TPA (tetradecanoyl phorbol acetate), a confusing abbreviation, since tPA is also the abbreviation of the serine protease Tissue Plasminogen Activator. The abbreviation PMA will therefore be used throughout this volume.

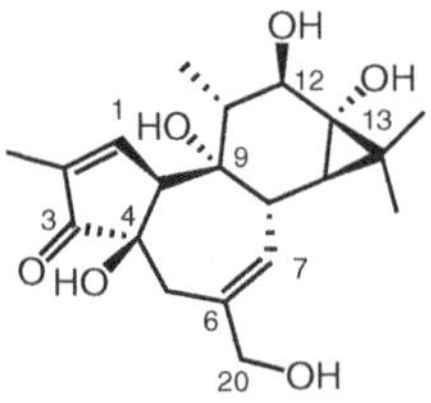

1 (phorbol)

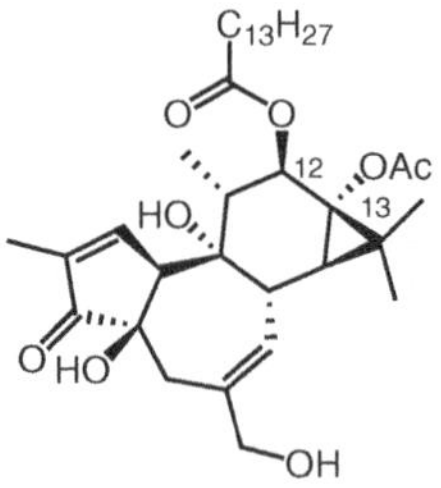

2 (phorbol myristate acetate)

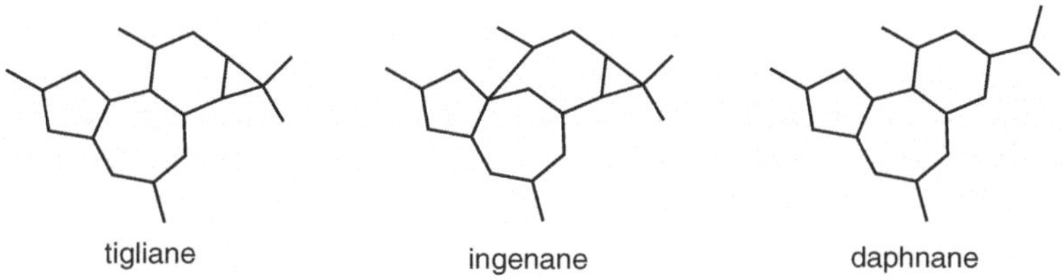

Fig. 1 The three basic phorboid skeleta

Tiglianes are narrowly distributed tetracyclic diterpenoids related in terms of biogenesis and constitution to ingenanes, daphnanes, and rhamnopholanes (Fig. 1). Overall, these four classes of diterpenoids are referred to as phorboids from the name of their biogenetic founding father. Phorbol (**1**) was obtained in pure form in 1927 [2] after a century of confusing studies on the nature of the irritant constituent of croton oil, and its structure was clarified by X-ray crystallography forty years later [3, 4] in the wake of a long series of constitutional and configurational revisions associated with attempts to elucidate its structure by a classic strategy based on degradation reactions. Studies on **1** were fueled by the discovery that its diesters are responsible for the tumor-promoting properties of croton oil [5], and were pursued systematically by Erich Hecker in Heidelberg and Fred J. Evans in London.

The clinical translation of tumor-promotion has been questioned and might well have been overemphasized [6], but the identification of protein kinase C (PKC) as a phorbol ester receptor [7] gave additional impetus to studies in this area, and the recent approval of an epoxytigliane ester (tigilanol tiglate) (**3**) [8] as a veterinary anticancer drug will undoubtedly continue to increase attention on this area of natural products research. Additional interest in the phorboids has come from research on oleaginous euphorbiaceous plants of economic relevance for the production of biodiesel fuel (*Euphorbia lathyris* L., *Jatropha curcas* L.) or for the preparation of bio-polymers (*Euphorbia lagascae* Spreng.), since the production of toxic phorboids interferes with their mass cultivation, processing, and use of by-products as animal feed [9, 10].

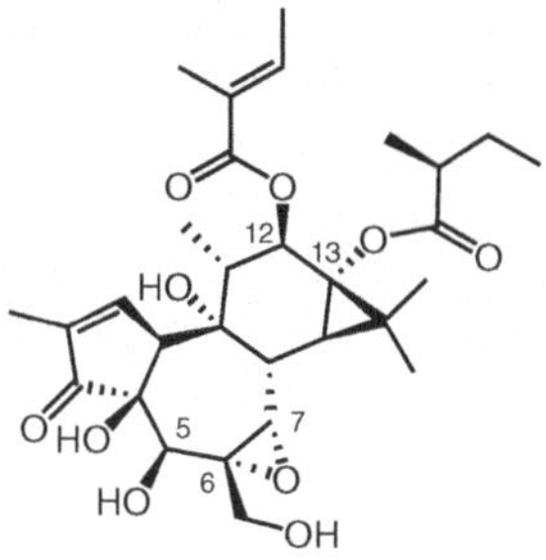

3 (tigilanol tiglate)

The present contribution covers in detail the distribution, chemistry, and molecular bioactivity of tiglianes from the very beginning of the studies on these compounds through to December 2023. Given its vast nature, the clinical and toxicological literature will be summarized mainly in terms of more recent and controversial aspects.

2 Historical Outline

The study of tiglianes is strictly connected to that of its archetypical member, phorbol (**1**), which is in turn related to the identification of the irritant principle(s) of croton oil. This extraordinarily irritant and obnoxious fixed oil is obtained from the seeds of *Croton tiglium* L. (Plate 1), a euphorbiaceous small tree native to India, Malaysia, and Indonesia. The poisonous properties of the seeds were documented in the medical literature of India and China, where this plant is known as Jamalgota, Mukula (India), and Ba-dou (China). In the Ayurvedic tradition, the seeds were not used as such, but only after detoxification by prolonged (3 h) boiling in cow's milk, a practice that considerably reduces the concentration of the irritant active constituents by partial extraction into the milk lipo-caseinic fraction and by thermal degradation [11]. In traditional Chinese medicine, *Croton tiglium* was considered one of the five supreme poisons, with the other four being *Aconitum lycoctonum*, *Veratrum* sp. v., *Aconitum fischeri*, and the Cantharides [12].

In addition to phorbol esters, the seeds of *C. tiglium* also contain crotin, a very toxic ribosome-inactivating protein rivaling ricin in terms of its LD_{50} (50 μg/kg in rabbits) [13]. Crotin, a mixture of two related toxalbumins, is suspected, along with ricin and abrin, to have been used to assassinate in 1978 the Bulgarian writer and journalist Georgi Markov in London, using an umbrella rigged to inject a poisonous pellet [14].

The first references to croton oil in the Arabic medical literature date from the thirteenth century, and the first mention in the European literature appeared in the "Tractado de las drogas y medicinas de las Indias orientales", an inventory of useful plants written in 1578 by the Portuguese naturalist and physician Cristóbal Acosta, who named the seeds of *C. tiglium* "Pinones de Maluco" (Fig. 2) [12, 15]. The medicinal uses of the oil were described in the systematic ethnopharmacological studies of Rheede ("Hortus Malabaricus", 1678) and Rumpf, whose 42-year long observations on the use of plants in the Indonesian island of Amboina led to him being accorded the title of "Plinius Indicus". In his "Herbarium Amboinense", published posthumously in 1743 after an adventurous journey of the manuscript from Ambona to Amsterdam, Rumpf wrote extensively on *C. tiglium*, describing the use of a drop of its seed oil in a glass of wine as a very powerful laxative [15]. The existence of a sort of ultra-potent version of castor oil took the fancy of European physicians, but the use of croton oil for obstinate constipation led also to dramatic cases of poisoning and was initially discouraged [12]. However, after a dosage readjustment (from half a drop to a maximum of two drops, with 20 drops being generally considered lethal),

Plate 1 *Croton tiglium* L. Photograph courtesy Vinayaraj; Creative Commons 3.0

croton oil became widely accepted as a drastic purgative, eventually being included in the 1827 Edition of the Prussian Pharmacopoeia ("Preussische Pharmakopöe. Fünfte Ausgabe"), and then in all other European pharmacopoeias [12].

Unusual formulations became popular, like the "pomum catharticum", prepared by dipping a lemon or a bitter orange in croton oil for a few weeks and then sprinkling it with sandalwood [12]. Croton oil was supposed to be quickly absorbed from the skin and, after dilution with olive oil, a tiny amount of it was rubbed on the belly skin of children to facilitate evacuation and diuresis [12]. When suitably diluted and applied as a liniment, croton oil was employed, in addition to denaturated ethanol, to relieve inflammation and rheumatic pain, and its use as a drastic laxative and as a counter-irritant was promoted by prominent clinicians all over Europe in both human and veterinary medicine [12].

In the first half of the nineteenth century, the burning oils of *Croton tiglium* and *Jatropha curcas* were often confused, even though their origin was different (Far East and South America, respectively). Thus, Pelletier and Caventou (who discovered quinine) carried out the first investigation on the nature of the irritant constituents of croton oil, but referred their observations to jatropha oil, promptly correcting the error in a subsequent publication [16]. In those times, sensory assays, and an alchemic heritage, dominated the quality control of medicinal plants, and were poorly discriminating. Thus, if the adulteration of croton oil with jatropha oil at least retained the biological and sensory profile of the medicine, that of bitter angostura (*Angostura*

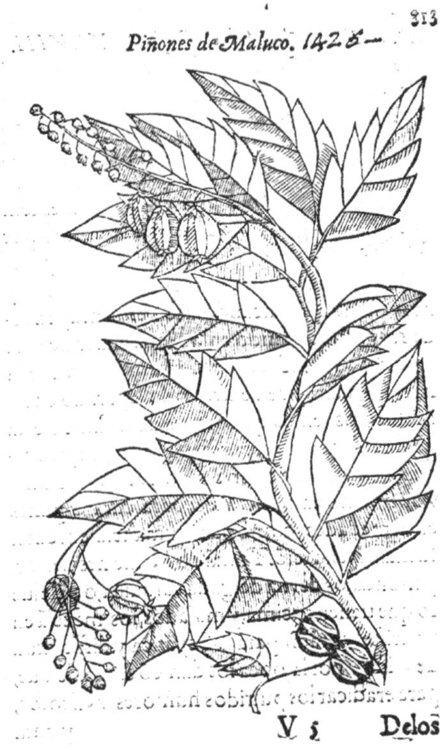

Fig. 2 *Croton tiglium* L. (Pinones de Maluco) as depicted in Cristóbal Acosta's Tractado de las drogas, y medicinas de las Indias Orientales (1578) (https://archive.org/details/tractadodelasdr00acosgoog/page/314/mode/2up?view=theater)

trifoliata (Willd.) T.S. Elias) with the bitter bark of the strychnine tree (*Strychnos nux vomica* L.) caused fatalities all over Europe [17].

Despite the technological poverty of early nineteenth century chemistry, some important observations were made on croton oil by using oral pungency as the end point. Thus, by extraction of the oil with aqueous ethanol, a resinous and irritant fraction could be separated in ca. 5% yield from an almost non-irritant fatty matrix. This, after treatment with diluted bases, however, became highly irritant, while strong bases completely destroyed the irritancy of croton oil and its resinous fraction [18]. These observations suggested that unlike castor oil, the irritant principle of croton oil was not a fatty acid, but a more polar compound, which was base-sensitive, and contained in a cryptic state in the fatty fraction of the oil, from which it was activated by a mild basic treatment.

Further advances were achieved in the second half of the nineteenth century, establishing that the major constituents of the lipid and the resinous fractions of the oil were esters of two distinct polyols, glycerol in the lipid phase, and an optically active unknown polyol in the resin fraction. This was the state of knowledge when Rudolf Böhm (1844–1926) started his studies on croton oil in Leipzig. Böhm was a leading figure in pharmacognosy at the turn of the nineteenth century, and conducted monumental studies on plant poisons (curare, muscarine, aconitine, veratrine). In

1923, he obtained the polyol in pure form from the resin fraction of croton oil [2], and established that the loss of activity of croton oil upon hydrogenation was due to the addition of hydrogen to this polyol [18]. The date of 1923 for the isolation of phorbol (**1**) was quoted by his collaborators (Böhm passed away in 1926) first in a PhD thesis [2] and then in a paper published in 1930 [18]. This makes the exact date of the isolation of phorbol (**1**) unclear, although it is referred to as 1923 in this article that appeared seven years after this feat. Böhm's collaborator Flaschenträger named the polyol phorbol (**1**) [19], possibly to avoid confusion with other names already derived from the root "croton" (crotonic acid, crotonic aldehyde, crotonic alcohol, crotyl radical). The suffix "ol" can be associated with the presence of hydroxy functions, and the prefix "phorb" to the botanical family of the Euphorbiaceae, to which *C. tiglium* belongs. Böhm also established a quantitative measure for the toxicity of croton oil based on its lethality to tadpoles (complete at a dilution of $1{:}10^8$ in the medium) and in frogs (lethal dose: 20 μg, sc) [15]. Research from the Leipzig group of Böhm led to the establishment of the correct molecular formula of phorbol (**1**) ($C_{20}H_{28}O_6$), identifying a polycyclic mono-unsaturated (actually doubly unsaturated) core, and the presence of five hydroxy groups, for which their types could not be clarified completely [19]. Of these, two were clearly tertiary and one primary, but the nature of the other two remained ambiguous. Phorbol (**1**) was non-pungent, but its per-esterification with acetic acid or higher homologues led to compounds active in toxicity studies, presumably due to hydrolysis of the primary 20-ester group in the water medium of the tadpole assay.

Working in Böhm's group, Flaschenträger discovered the rearrangement of phorbol (**1**) to crotophorbolone (**4**), a reaction of critical relevance for the structure elucidation of **1** [19]. In addition to the structure of phorbol, a second unresolved issue was the identification of its active form in the resinous fraction of croton oil, a challenge for which the relevance was stressed by Berenblum's discovery of the "co-carcinogenic" properties of croton oil in 1941 at Oxford University [20]. Surprisingly, this discovery had no effect on the growing popularity of lay face peelers, with various croton oil-phenol skin rejuvenating formulas having been used since the 1920s in Hollywood by many, if not most, of the leading "stars" of the day and, possibly, by millions of people as well, if this estimate is indeed correct [21].

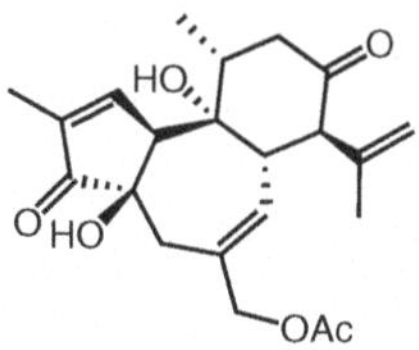

4 (crotophorbolone)

Capitalizing on the development of column chromatography and countercurrent fractionation, native phorbol esters eventually were able to be obtained from croton oil in the 1960s. Alumina adsorption chromatography made it possible to separate the resinous fraction into several subfractions having different inflammatory and vesicant properties, while countercourrent chromatography was used to sort out the most active resin fraction into two TLC-pure compounds, named compounds A and compound B, both highly irritant. By further tuning the composition of the biphasic countercurrent system, compounds A and B were shown by Hecker to be a mixture respectively, of four (A_1 to A_4) and seven (B_1 to B_7) compounds [22].

In the meantime, significant progress had also been made with the structure elucidation of phorbol (**1**). In the late fifties, Kauffmann clarified the nature of the two "ambiguous" hydroxy groups, and showed the presence of enone, glycol, and cyclopropane functionalities [23]. However, the propensity of phorbol to undergo skeletal rearrangements made it impossible to elucidate its structure with classic degradation studies, given the unreliable reactivity of some of its functional groups. In addition, the tendency of phorbol (**1**) to form solvates with practically all solvents was also shared by some of its derivatives, making the determination of their molecular formulas difficult. An especially misleading claim by Kauffmann was that treatment of **1** with lead tetracetate generated one molecule of acetone, implying the presence of a cleavable glycol system [24].

Tetracyclic structural proposals for phorbol (**1**) were first made in 1965 by Arroyo [25] and by Hecker [26]. The two formulas **5** and **6**, drawn without stereochemical features, differ in the location of the enone system, situated on a 5-membered ring by Arroyo and on a 7-membered ring by Hecker, who in 1966 provided additional spectroscopic evidence for formula **6** [27].

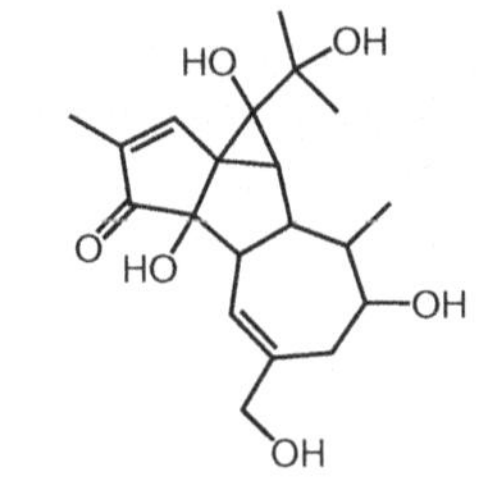

5

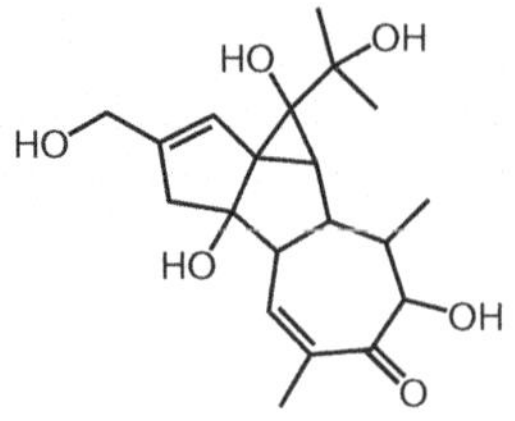

6

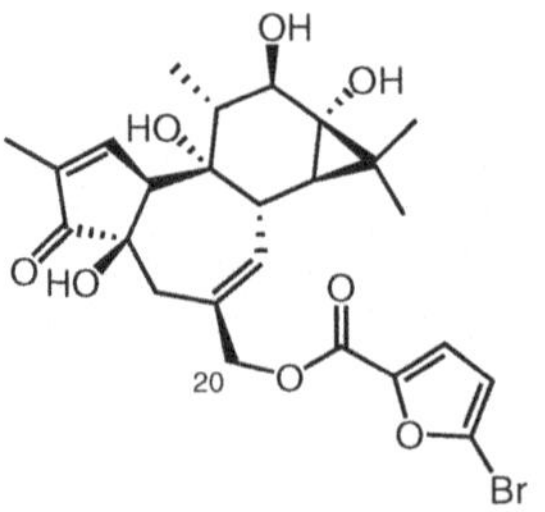

7 (phorbol 20-(5-bromo)furoate)

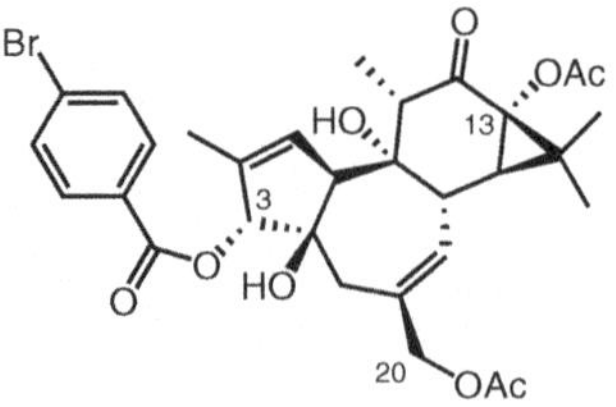

8 (3-*p*-bromobenzoate of neophorbol 3,20-diacetate)

Eventually, in 1967 the correct structure of phorbol (**1**) was proposed independently by Crombie and Hecker based on the X-ray analysis of the chloroform solvate of phorbol 20-(5-bromo)furoate (**7**) [3], and of the 3-*p*-bromobenzoate of neophorbol (= 12-dehydro-3αH-dihydrophorbol) 3,20-diacetate (**8**), respectively [4]. Owing to the systematic studies by Hecker on tigliane diterpenoids, the contributions of Crombie to the structure elucidation of phorbol (**1**) surprisingly have been underestimated. After establishing the structure of the natural product, Crombie, who in 1969 moved from Cardiff to Nottingham, abandoned this area of investigation, making impressive contributions (inter alia) to the chemistry of cannabinoids, steroids, pyrethroids, and Khat alkaloids, discovering avenacins and also reporting several total syntheses of natural products, including the tigliane precursor casbene [28]. Conversely, Hecker dedicated most of his scientific life to the study of the phytochemistry and biology of phorboids, authoring over 300 articles in this area and forever associating his name with these compounds.

The issue of priority in the structure elucidation of phorbol (**1**) is controversial and deserves a comment. Simultaneous discoveries are by no means rare in science. The discovery of oxygen by Scheele, Priestley, and Lavoisier is a famous example, and, just like the race to synthesize Taxol, also that leading to the structure elucidation of **1** ended up in a substantial tie. The date of publication of the structure elucidation of phorbol would prioritize Crombie, just like, based on laboratory book dates, Scheele would be prioritized for the discovery of oxygen. However, it was Lavoisier who capitalized on the discovery of oxygen to progress science, and it was Hecker who furthered the phytochemical and biological implications of the structure elucidation of phorbol (**1**), casting, accordingly in the role of Lavoisier in also emphasizing his own contributions. The final events commence in 1965 with the correction by Hecker, in light of a careful analysis of 60 MHz ^{1}H NMR data [26], of the tetracyclic structure **5** suggested by Arroyo the same year [25]. The revised structure **6** proposed by Hecker was in accordance with the spectroscopic properties expected for the structural elements identified in the degradation experiments, also including the alleged production of acetone upon treatment with lead tetraacetate. The limited spectroscopic capabilities of the time did not allow for the unambigous establishment of a compound under study, in the same way the structural elements identified by chemical reactivity and UV/IR spectroscopy were assembled. However, the formula **5** proposed by Arroyo was closer to the correct one for **1** than that by Hecker.

Both structures were questioned by Crombie, who found that the oxidative degradation of phorbol did not actually produce significant amounts of acetone (see Sect. 4 for a discussion of this critical reaction). Given the instrumental limitations of NMR spectroscopy in the 1960s, the resolution of the structure of phorbol (**1**) could only be settled by X-ray analysis. Crombie's data on the chloroform solvate of a phorbol 20-ester (5-bromofuroate) (**7**) were the first to be published in 1967 [3], and established the connectivity and relative configuration of the natural product. The constitution was in agreement with a suggestion Hecker had proposed in a lecture at the Chester Beatty Research Institute in London on 11th January, 1967, as acknowledged by Crombie in his article, which was received by Chemical Communications on June 9th of that year [3]. The crystallographic data on the chloroform solvate were collected at a temperature of −160°C to prevent desolvation and destruction of the crystalline framework. Hecker's data on neophorbol 12,13-diacetate-3-*p*-bromobenzoate (**8**) appeared later that year in a paper published in September [29]. Given the presence of a carbonyl function at C-12 in neophorbol, no information was provided on the configuration of the 12-hydroxy group in the natural product, but this revised the absolute configuration Hecker had previously assigned based on circular dichroism data [29].

In conclusion, both Hecker and Crombie significantly contributed to the clarification of the structure of phorbol, which should be credited to both authors, and not only to Hecker, as found in the literature. Based on the published evidence, Hecker first presented the correct constitution of phorbol (**1**) in public seminars, but Crombie was the first to publish evidence for it and to establish the relative configuration of the natural product. In addition, Crombie's full paper on phorbol published in 1968 [30] summarizes the whole early chemistry of this compound, and, by providing detailed experimental descriptions for all reactions, compensated for the lack of full experimental protocols that plagued the early literature on phorbol (**1**).

In light of the structure elucidation of phorbol (**1**), the difference between type-A and type-B phorbol esters was traced to the location of the long-chain acyl moiety that is bound to the secondary hydroxy group in type-A esters, and to the tertiary hydroxy group in type-B esters [31]. The second ester group was acetic acid in type-A esters, and a mixture of short-chain acids (acetic, tiglic, α-methylbutyric) in type-B esters. The most potent and major irritant principle of croton oil turned out to be factor-A_1, phorbol 12-myristate-13-acetate (**2**), which could fully reproduce the same biological profile as croton oil in tumor-promotion assays [31].

These structural studies paved the way to the identification of a phorbol ester receptor, eventually identified in 1984 by Nishizuka as a series of isoforms of protein kinase C (PKC) as well as of an endogenous biological analogue, diacylglycerol (DAG) [7]. The protein kinase C family of serine/threonine kinase is involved in the regulation of a vast array of cellular functions, and, because of its kinase-activating properties, phorbol 12-myristate-13-acetate (**2**) has become an indispensable tool to study cell function, and also is currently undergoing clinical development as an anticancer agent (see Sect. 6.3).

Using in vivo biological assays (mouse ear erythema assay, tumor promotion in mice) and countercurrent chromatography, Hecker characterized several inactive analogues of phorbol esters in croton oil [31]. He also postulated a triester structure for the cryptic phorbol esters remaining in the lipid phase after extraction of the oil with aqueous ethanol, suggesting that mild basic treatment removes the acyl group on the primary 20-hydroxy group, generating irritant diesters from the native triesters which, due to their lipophilicity, are not soluble in polar phases [31]. Following the structure elucidation of phorbol (**1**), in addition to the seminal studies by Hecker, significant contributions also came also from Fred Evans at the University of London School of Pharmacy. Hecker and Evans independently wrote the first two comprehensive reviews on phorbol and related compounds, which appeared in volumes 31 [32] and 44 [33] of this book series, in 1974 and 1983, respectively.

During the past four decades, progress from these seminal contributions was basically incremental as regards phytochemistry, since the hundreds of new members added to the tigliane inventory were mostly variations of chemotypes already described in earlier studies. Conversely, dramatic progress was made in the study of the reactivity (see Sect. 4) and the total synthesis (see Sect. 5) of tigliane esters, for which their bioactivity focus shifted from the area of tumor promotion to tumor suppression (see Sect. 6.3). Interest in the clinical potential of tigliane derivatives has, in fact, long lagged behind those of other phorboids, such as the ingenane ingenol-3-angelate (**9**) (Picato®) [34] and the daphnane resiniferatoxin (**10**) (RTX) [35]. However, this was revitalized by the approval of the tigliane, tigilanol tiglate (**3**) [8], as a veterinary anticancer agent by the EMA in Europe and the FDA in the U.S. The potential beneficial use of this compound also for the management of human malignancies, and the growing evidence for the anticancer activity of PKC [6], have combined to bring tigliane diterpenoids under the spotlight of anticancer research.

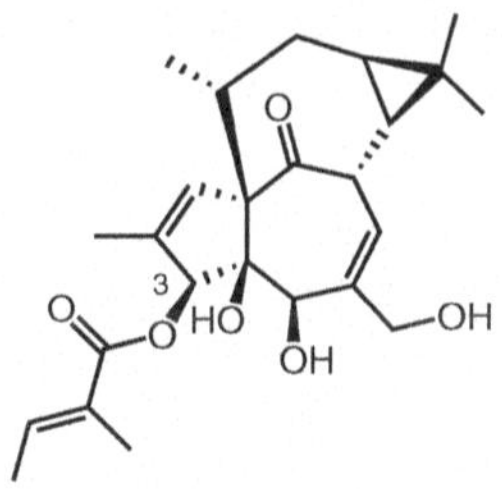

9 (ingenol-3-angelate)
an ingenane

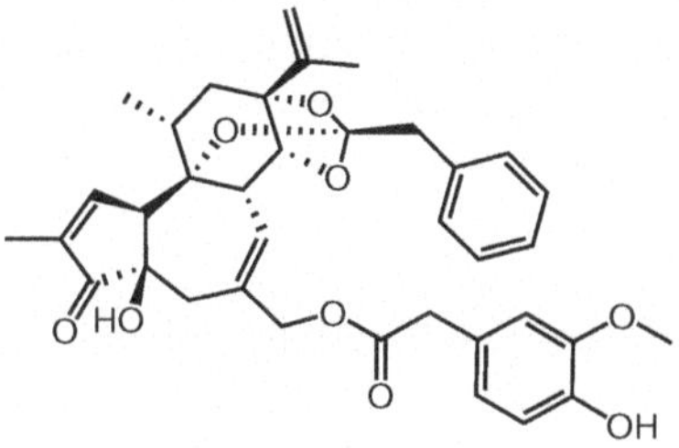

10 (resiniferatoxin)
a daphnane

3 Phytochemistry

3.1 Isolation

Tigliane diterpenoids occur as complex mixtures of esters and have never been reported as free polyols from hydrolytically unprocessed or otherwise degraded extracts. Depending on the degree of esterification and on the nature of the acyl moieties, they show different levels of lipophilicity, peaking with long-chain triesters and gradually decreasing with the number of acyl moieties, their shortening, and the introduction of polar groups, mainly basic amino groups. Tigliane 12,13,20-triesters are "cryptic", since they do not show activity in the classic bioactivity assays for phorbol esters (i.e. in tumor promotion and the mouse ear erythema assays), and cannot be efficiently recovered from the fatty matrices of euphorbiaceous seeds [32]. Due to of the difficulty of purification, their characterization has only been achieved recently, capitalizing on their occurrence in a lipophilic matrix made up not by esters, but by fatty acids, which could be removed by extraction with bases [36]. While acylation is universal within the inventory of natural conventional tiglianes, glycosidation is very rare, and represents a veritable type of apex in the inventory of this class of diterpenoids (see Sect. 3.2).

The deconvolution of complex tigliane mixtures is greatly helped by HPLC-MS-MS analysis, since they are generally based on a dominant, or even single, polyol, making it possible to identify their acyl moieties based on a diagnostic fragmentation pattern (see Sect. 3.2) [36]. A combination of normal-phase and reversed-phase solid-liquid chromatographic techniques (gravity column chromatography, vacuum chromatography, preparative HPLC) as well as liquid-liquid partition techniques (rotation locular- and droplet countercurrent chromatography) have been employed successfully to isolate single tigliane esters [37]. These techniques have substantially replaced the liquid-liquid distribution methods (O'Keefe and Craig distribution), which were critical in isolating the first native tigliane esters from croton oil [32]. The chromatographic protocols used to isolate tigliane esters are mainstream in phytochemistry, and will not be discussed further. Conversely, the isolation of the main tigliane polyols has faced considerable challenges, owing to their instability and solubility in water.

Isolation of the Basic Tigliane Polyols

Suitable sources for three basic tigliane polyols can be identified within plants cultivated for medicinal or technical purposes, and therefore potentially available in large amounts to sustain medicinal chemistry campaigns: the seeds of *Croton tiglium* for phorbol (**1**) [32], the kernels of *Fontainea picrosperma* C.T. White for 5β-hydroxy-6α-epoxyphorbol (**11**) [8] and the leaves and seeds of *Aleurites fordii* Hemsl. (tung oil plant) for 16-hydroxyphorbol (**12**) [38]. No convenient source of 12-deoxyphorbol (**13**), the most widespread tigliane polyol is known, since its esters do not seem to be accumulated in significant amounts in any of their numerous plant sources. However, an ingenious protocol has been developed for the preparation of its 13-esters by Wender (see Sect. 4.3).

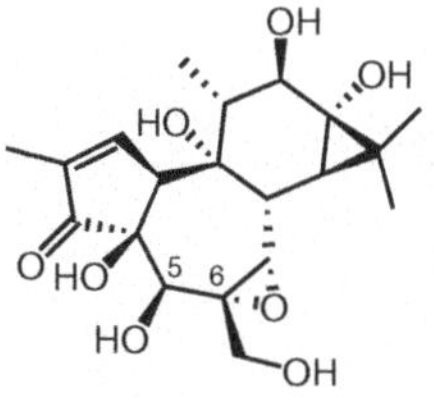

11 (5β-hydroxy-6α-epoxyphorbol)

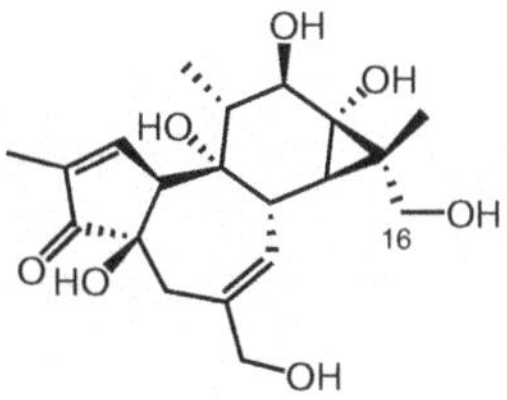

12 (16-hydroxyphorbol)

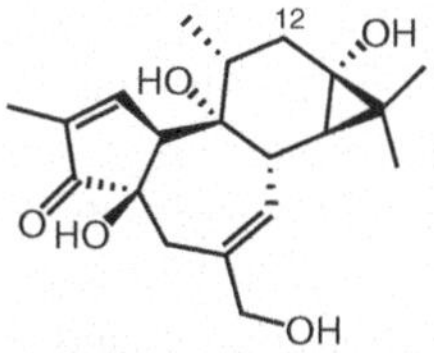

13 (12-deoxyphorbol)

Phorbol (**1**) and, to a minor extent, 12-deoxyphorbol (**13**) are stable during the basic conditions required for the removal of their ester decoration, while the other relevant polyols are degraded or epimerized during hydrolytic and/or transesterification conditions. Protection of the 5,20-hydroxy groups as an acetonide stabilizes 5β-hydroxy-6α-epoxyphorbol (**11**) toward bases [8], while 16-hydroxyphorbol (**12**) has so far proved elusive, since hydrolysis of its esters was accompanied by extensive degradation, and produced only the corresponding 4α-derivative of the basic polyol as the sole recovered tigliane [38]. The degradation of 16-hydroxyphorbol (**12**) could have an interesting biogenetic relevance, since under acidic conditions, it could trigger rearrangement of the tigliane to the daphnane skeleton, as discussed in Sect. 3.4.

The Isolation of Phorbol

Isolation from croton oil has been the most common way to obtain phorbol (**1**). However, owing to its high oleic acid content (>50%), croton oil has also raised interest as a starting material for the production of biodiesel. This apparently has led to a significant reduction in its biological potency, since varieties with low levels of tigliane esters might have been selected and cultivated [39]. Alternatively, the commericial oil could have been heated to decrease the titer of the irritant phorbol esters, while an increased content of α-branched esters has also been observed in various commercial samples (type-C phorbol esters). These branched esters require strong basic treatment to remove their acyl decoration (see below), during which significant amounts of 4α-phorbol (**14**) are formed, with an overall decrease of the final yield of phorbol (**1**) [32].

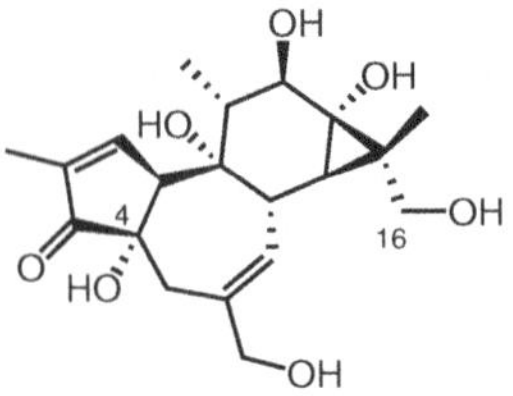

14 (4α-phorbol)

The main difficulties in the isolation of **1** from croton oil or from croton seeds are associated with the water solubility and the instability of this polyol, and with the presence of triesters, which, due to a very high lipophilicity, are not recovered well from a fatty matrix (cryptic esters). Owing to its water solubility, phorbol partitions with glycerol in the workflow of the oil hydrolysate, and the presence of glycerol makes the crystallization of phorbol difficult and erratic. In addition, some solvates, like that with ethanol, have a paradoxical limited stability despite their highly crystalline state [32]. In general, amorphous phorbol strongly retains all solvents used for its chromatographic purification, but shows a higher stability compared to some crystalline solvates. It is therefore advisable to crystallize **1** only for analytical purposes like the registration of its melting point and optical rotation, which should, in any case, be different for distinct solvates, and also the compound is better stored as a powder or as a foam. Phorbol (**1**) has a strong tendency to retain all types of solvents [32], and the unavoidable presence of alcohols means that an excess of acylating reagents should be used for the synthesis of targeted esters. When high quality croton oil is available, obtaining **1** is easier than from the seeds, and avoids exposure to the toxic proteinaceous constituents of the plant.

Phorbol from Croton Oil

Croton oil is commerically available, although its price has increased significantly over the past few years, being currently available from Sigma at € 250/100 cm^3 [40]. The growing popularity of the croton oil peel [21] could also have been a factor, but the massive cultivation of *Croton tiglium* for the production of biodiesel [39] makes such a high price surprising. Four different isolation protocols are presented, differing for the stage of the deacylation, its hydrolytic or transesterificative nature, and the purification strategies of the crude deacylated mixture. In all cases, the isolation yield was reported in the range of 1%, but the current quality of croton oil is highly variable, and it is therefore advisable to check on a small scale its suitability for the isolation of phorbol (**1**) before scaling up the process. All three protocols are variations on the original Boehm-Flaschenträger process [18, 41].

(a) Hecker [32]: Croton oil (500 cm^3) is dissolved in 2% methanolic barium hydroxide heptahydrate (2.25 dm^3), and the solution is stirred under a nitrogen atmosphere for 10–12 h. The precipitate of barium carboxylates is removed by filtration, and the basic filtrate is evaporated to dryness and next taken up in water (2 dm^3) and extracted with ether. The water phase is acidified to pH ca.

5 with 2 *N* H_2SO_4, 40 cm^3 of sat. sodium sulfate are added, and the solution is next stored at refrigerator temperature (4°C) for 12 h. The copious precipitate of barium sulfate is filtered off, and the pH of the filtrate is brought to ca. 7 with 2 *N* sodium hydroxide. The neutralized water solution is extracted sequentially with ethyl acetate (500 cm^3) and ether (500 cm^3), evaporated to dryness, and then digested with ethanol (100 cm^3) to precipitate sodium sulfate. After filtration, the ethanol solution is concentred to ca. half volume and stored at 4°C to induce the precipitation of phorbol (**1**) as an ethanol solvate, a process completed in about four weeks. The solvate (5.5–6 g) is not stable and, after drying, is dissolved in warm (ca. 60°C) water. The solution is concentrated until a precipitate starts forming, and is then stored at 4°C for one week to complete the process. Overall, some 4.4 g (ca. 0.9%) of phorbol (**1**) are obtained as a white powder, having a mp of 250–251°C.

This protocol is laborious and lengthy, but minimizes the contact with croton oil by the operator, which is treated directly with bases and therefore biologically "neutralized" at the very outset of the process. On the other hand, the precipitation of **1** from a viscous glycero-ethanolic matrix is highly erratic, and, at least in the experience of the present authors, difficult to reproduce and standardize. In addition, the instability of the ethanol solvate requires an additional step to convert **1** into a stable hydrate.

(b) Crombie [30]: This protocol starts with the multiple (fourfold) extraction of croton oil (2 kg) with methanol (overall 6.3 dm^3). Each time the mixture is stirred for several hours (an overall 15 h), and then set aside to promote phase separation. The pooled upper methanol phases are brought to pH ca. 9 with 1 *N* methanolic barium hydroxide (395 cm^3). After two days, due to a pH decrease, an additional 10 cm^3 aliquot of the basic solution is added, and the hydrolysis is continued for an additional two to three days. The orange-red methanol solution is evaporated and the residue (192 g) partitioned between water (1.5 dm^3) and ether (1.1 dm^3). The water phase is then treated with sodium sulfate to remove barium as its insoluble sulfate, and the filtrate is acidified to ca. pH 5 with 2 *N* sulfuric acid, and sequentially extracted with ether (2 × 150 cm^3), ethyl acetate (2 × 100 cm^3), and ether (100 cm^3; presumably to partially remove ethyl acetate from the water phase). The water solution is neutralized with 1 *N* sodium hydroxide and evaporated, and the residue is treated with hot methanol (600 cm^3) to separate phorbol from insoluble inorganic salts and other impurities. Removal of the solvent gives a gum (68 g) that is crystallized from methanol/ethyl acetate to afford four successive crops of phorbol (**1**) (overall yield of **1**, 23.6 g, ca. 1.1%).

The present authors found this protocol reproducible, but dangerous because of the initial extraction of the oil with methanol, and a need for the manipulation of a concentrated solution of the highly irritant and toxic phorbol esters. In addition, cryptic esters are only slowly and erratically extracted from the oil with methanol, requiring long stirring times and intervals to allow phase separation.

(c) Appendino et al. [42]: this protocol was elaborated during more than two decades of experiments and progressive improvements in various details, and differs from the previous methods because phorbol (**1**) is released from its esters by direct transesterification of the oil with sodium methoxide, a maneuver that minimizes the handling of croton oil and also fully recovers the cryptic esters. In turn, the separation of phorbol from glycerol is accomplished by partition between tetrahydrofuran and water, capitalizing on the complementary solubility of phorbol in tetrahydrofuran and glycerol in water. Croton oil (500 cm^3) is measured in a graduated cylinder and poured into a 2 dm^3 round-bottomed flask, washing the graduate cylinder and the funnel with methanol (50 cm^3). Freshly prepared 0.3 *N* sodium methylate (ca. 1 dm^3) is then added slowly, until a pH of 12–12.5 is reached. The solution is stirred at room temperature, monitoring the course of the transesterification by TLC and checking the decrease of the pH, which after one to two hours stabilizes at around 11. After ca. 12 h, the reaction, now biologically "deactivated", is transferred into a separation funnel, neutralized with a few drops of glacial acetic acid, and extracted with petroleum ether to remove the fatty acid methyl esters. Evaporation of the methanol phase affords a black resinous oil, which is dissolved in THF (250 cm^3) and washed with brine to remove glycerol (5×250 cm^3). The combined brine phases are counter-extracted with THF, and the pooled organic phases are dried (Na_2SO_4) and evaporated, obtaining a gummy residue (ca. 30 g). This is purified by gravity column chromatography on silica gel (250 g) with a petroleum ether-ethyl acetate gradient. Fractions eluted with petroleum ether/ethyl acetate (2/8) afford a mixture (ca. 6 g) of phorbol 12-monoesters with short chain α-branched acids (tigliate, α-methylbutyrate). After their complete elution, the mobile phase is changed to ethyl acetate/methanol (9/1), eluting 15 g of a ca. 5/1 mixture of **1** and 4*α*-4-deoxyphorbol. Crystallization from ethyl acetate affords **1** (ca. 6 g, 1.2%) as an oatmeal-colored amorphous powder, devoid of its 4-deoxy derivative and sufficiently pure to be used in synthetic projects. For longer storage, it is advisable to crystallize the ethyl acetate solvate from methanol, obtaining a solvate stable for years at low temperature (4°C) and in a closed container protected from light. The mixture of α-branched monoesters obtained as side-products could not be converted to phorbol without significant isomerization to 4*α*-phorbol (**14**).

(d) Xie [43]: A streamlined combination of hydrolysis and extraction was developed to minimize the exposure to the phorbol diesters contained in the oil. Thus, the K_2CO_3-promoted transesterification of glycerides and diterpenoid esters from croton oil was faster in a biphasic methanol-petroleum ether system compared to a monophasic methanol solution. The detrimental effect on the rate of transesterification by fatty acid methyl esters could simply be due to a mass transfer effect associated with the partition of the methyl esters into the petroleum ether phase, with consequent escape from the equilibrium that is established in the methanol phase, where the reaction takes place and where glycerol and phorbol are, conversely, well soluble. Thus, one gram of croton oil is dissolved in 60 cm^3 petroleum ether, and then 15 cm^3 of methanol containing 450 mg K_2CO_3 are added. The biphasic system is stirred vigorously for 90 min at 20°C, the phases are separated, and a phorbol-containing fraction is recovered by evaporation of the methanol phase. The residue can be purified directly by high-speed counter-current chromatography using as stationary phase the upper phase of the ternary mixture ethyl acetate/*n*-butanol/water 4.7/0.3/5 (*v*/*v*/*v*) containing 0.36 g Na_2SO_4/10 cm^3. The diterpenoid fraction obtained with this protocol is a mixture of five compounds, but only phorbol could be characterized, since two constituents degraded during their recovery from the eluent solution, and the other two could not be separated.

In the lack of precise yield data, it is difficult to judge the value of the Xie protocol, but the biphasic hydrolysis is, undoubtedly, interesting, as would be the identification of the by-products associated with this process. A further issue worth evaluating is the recovery of phorbol triesters, since they partition selectively into the lipophilic phase, although the facile loss of their 20-acyl moiety should guarantee their funneling into the hydrolytic process and eventually lead to complete transesterification.

Phorbol from *Croton tiglium* Seeds

Owing to the declining quality of commercially available croton oil in terms of phorbol (**1**) content, two groups (Christensen and Wender) have described in detail the isolation of this polyol from croton seeds [44, 45]. The continued medicinal uses for the seeds in folk medicine and in homeopathy apparently has preserved their phytochemical quality throughout the supply chain. The isolation of phorbol (**1**) from croton seeds requires great care to avoid contamination not only from the phorbol esters present, but also from the highly toxic protein crotin. This ribosome-inactivating phytalbumin is related to ricin from castor seeds, and no antidote is known against these types of ribosome inhibitors [46]. The absorption from the intact skin is presumably very poor, but, since castor seeds used in necklaces have been reported to cause skin reactions possibly related to their content of ricin [47], absorption of crotin through small skin scratches present on the skin of operators could not, in principle, be ruled out. The yields reported for the isolation of phorbol from croton seeds are only apparently lower than that from croton oil, being actually comparable if corrected for the oil content (ca. 20%) of the raw seeds.

(a) Christensen [44]: ground seeds (1.493 kg) are extracted by repeated sonication with EtOAc at 40°C (three times, for overall 6 h and 1.6 dm^3 of EtOAc). The combined extracts are concentrated, dissolved in methanol (0.5 dm^3), and shaken until two phases are formed. The upper (possibly lower, due to an error in the original article) methanol phase is separated, and the triacylglycerol-containing phase is washed with two additional portions of methanol (2 × 600 cm^3). The combined methanol phases are evaporated, and the residue is purified by gravity column chromatography on silica gel (600 g) using a heptane-EtOAc gradient. After a second chromatographic purification with a CH_2Cl_2-methanol gradient, a mixture (ca. 12 g) of phorbol esters is obtained. Part of this mixture (3.52 g) is hydrolyzed at 0°C with a solution of barium hydroxide heptahydrate (3.8 g) in methanol (80 cm^3). After 18 h, the reaction is worked up by neutralization with 2 *N* H_2SO_4, filtration, and evaporation. The residue is purified by gravity column chromatography on silica gel with a CH_2Cl_2-methanol gradient, obtaining 1.06 g (0.22%) of phorbol (**1**) from the fractions eluted with CH_2Cl_2-methanol 12:1.

(b) Wender [45]: seeds (5 kg) are powdered and Soxhlet-extracted in 500 g batches and 1 dm^3 methanol for 15 h. After evaporation, the combined extracts are suspended in methanol (2 dm^3) and transferred into a 4 dm^3 separatory funnel. The lower phase is washed repeatedly (3 × 200 cm^3) with methanol, and the pooled upper methanol phases are transesterified with caesium carbonate (140 g) at a pH of ca. 11. After 18 h, the reaction mixture is neutralized with 6 *N* H_2SO_4, bringing the pH to ca. 5.5. After filtration and concentration, the residue is dispersed in water, saturated with sodium chloride, washed with ether (3 × 500 cm^3), and then extracted with THF (5 × 500 cm^3). Evaporation of the combined THF extracts gives a mixture of a solid and an oil, which is then taken up in absolute ethanol and concentrated again to remove residual water. The residue is dispersed in methanol (200 cm^3), adsorbed on silica gel (200 g), and purified by silica gel vacuum column chromatography using a CH_2Cl_2-methanol gradient. Phorbol (**1**) (16.1 g, 0.32%) is obtained in this way as a white foam.

Isolation of Phorbol as Its 20-Trityl Ether

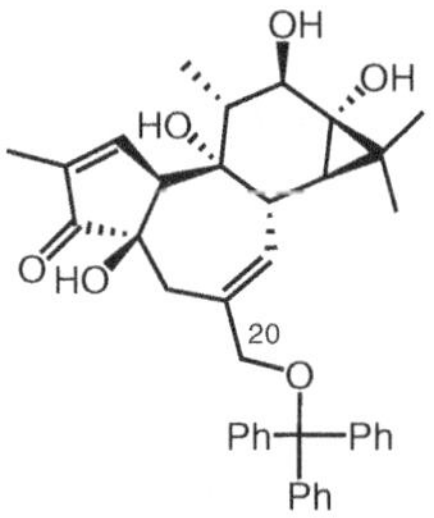

phorbol 20-trityl ether

The protocol was developed in Hecker's Heidelberg laboratory. It was privately communicated by Dr. Bernd Sorg to one of the present authors (GA) in 1992, and was published, presumably after approval of the originators, by Newton in 2003 as part of a paper on the structure-activity relationship of phorbol esters [48]. Phorbol 20-trityl ether is a very convenient storage form for a native polyol, in being stable for years at room temperature in closed vials. The bulky protection of the 20-hydroxy group prevents oxygen attack on the Δ^6-double bond, also increasing, for unclear reasons, the stability toward base-induced epimerization of the 4-hydroxy group [49]. This surprising bonus makes it possible to use more basic conditions to hydrolyze the native esters, recovering also those containing unsaturated (tigliate) or α-branched (α-methylbutyrate) esters, which are only slowly hydrolyzed under milder basic conditions. The initial acid-catalyzed transesterification removes chemoselectively the acyl moieties at the primary allylic 20-hydroxy group, securing full recovery of the cryptic triesters and setting the stage for the trityl protection. Under these conditions, triglycerides are transesterified, and can be removed by extraction with *n*-hexane. The protocol requires expertise in the manipulation of toxic compounds, since it involves the handling of a concentrated mixture of phorbol diesters, which are much more irritant than the starting oil. In addition, the viscous mixture of phorbol diesters obtained after the chemoselective transesterification strongly retains methanol, and a large excess of trityl chloride is therefore necessary to complete the protection.

(a) Original protocol: Croton oil (500 cm^3) is suspended in 10^{-2} *M* methanolic $HClO_4$ (1.75 dm^3), and mechanically stirred for 40 h at room temperature, resulting in the formation of a brown emulsion. After neutralization with solid NaOAc (5–8 g), depending on the acidity of the starting oil), the solution is extracted with hexane (3 × 1 dm^3) to remove fatty acid methyl esters, and then evaporated to obtain a brown syrup. The latter is taken up in pyridine (500 cm^3) and treated with trityl chloride (150 g). After stirring at room temp. for 72 h, the reaction is worked up by partition between EtOAc (800 cm^3) and sat. $NaHCO_3$ (800 cm^3). The organic phase is washed sequentially with sat. $NaHCO_3$ (2 × 400 cm^3) and with a 1/1 (*v*/*v*) mixture of brine and 1 *N* HCl (2 × 400 cm^3), and then evaporated. The residue is treated with methanol (ca. 500 cm^3) to precipitate triphenylmethanol, and solid NaOMe (ca. 4 g) is added to the filtrate. After stirring for 22 h at room temp., the reaction is neutralized with 10% methanolic acetic acid and evaporated. The residue is purified by gravity column chromatography on silica gel, using 10% acetone in CH_2Cl_2 to remove residual trityl alcohol and lipophilic impurities, and then 1:1 acetone-EtOAc to elute 7.4 g phorbol 20-trityl ether (1.5%) as a colorless foam.

(b) Modified protocol (Appendino et al., unpublished): the acidic transesterification of 100 cm^3 croton oil is carried out in a 1 dm^3 round-bottomed flask using 300 cm^3 of 10^{-1} *M* methanolic $HClO_4$. The suspension is stirred on a rotary evaporator for 24 h at the minimum speed and without applying vacuum. After neutralization with solid NaOAc (ca. 1 g), a vacuum is applied until no more methanol is condensed. The residue is taken up in hexanes, silica gel (150 g) is added, and the solvent is then evaporated. The resulting powder is transferred from the original reaction flask into a sintered glass filter where a bed of Celite and sodium sulfate has been deposited, and the cake is washed sequentially, applying a light vacuum, with hexanes (1 dm^3) to remove fatty esters, and next with CH_2Cl_2 to recover the phorbol diesters. The CH_2Cl_2 filtrate is collected in a round-bottomed flask, and, after the addition of 100 cm^3 toluene, the flask is directly connected to a rotary evaporator, and the solvents are removed at a temperature lower than 40°C. The alcohol- and water-free deep reddish oil remaining in the flask is then dissolved in pyridine, treated with trityl chloride (8 g), and then processed according to the original protocol.

This modification minimizes the handling of toxic and irritant diterpenoid esters by conducting various procedures within the same flask. Furthermore, it avoids the use of a large excess of trityl chloride, thus streamlining the final chromatographic purification of phorbol-20-trityl ether.

The Isolation of Cryptic Tigliane Esters

It has been known since the mid-1850s studies on croton oil by Rudolph Bucheim, the father of modern pharmacology, that the tasteless and non-irritant fraction remaining after exhaustive ethanol extraction of phorbol esters acquires both an irritancy and an extremely sharp taste after mild basic treatment [50], suggesting the presence in croton oil of cryptic, or non-extractable and non-irritant, forms of phorbol esters. Modern investigations have confirmed this observation, rationalizing it in terms of the presence of long-chain phorbol triesters not extractable or only partially selectively extractable from the lipid matrix, just like some classes of condensed tannins [51]. Their recovery could be easier from the plant latices, but data on the relative amounts of cryptic and extractable tiglianes are only available for the seeds of *Croton tiglium* [32] and for the kernels of *Fontainea picrosperma* C. T. White [8], the sources of, respectively, phorbol (**1**) and its 5β-hydroxy-6α-epoxy derivative (**11**). Based on the comparative extraction yield of phorbol from croton oil and from the diterpenoid esters recovered from the oil with alcohol extraction, a ratio of ca. 1:1 was estimated between cryptic and extractable phorbol esters [5], but other studies found the triesters to be more abundant than diesters, suggesting a ca. 4:1 ratio, with only about half of the triester fraction being recovered by extraction of croton oil with methanol [39]. The fatty matrix of *F. picrosperma* kernels is made up not by glycerides, but rather by fatty acids, and this makes possible their removal by extraction with bases, leaving in the organic phase the whole suite of tigliane esters [36]. From ^{1}H NMR measurements, a ratio of ca. 1:2 was derived between cryptic and "plain" esters, and the third esterification site was located at the 20-hydroxy group. The easy removal of the third acyl moiety in cryptic tiglianes from croton oil [32] strongly suggest a similar location for the third acyl moiety.

Leaves are not an accumulation organ for tigliane diterpenoids, but, owing to the ubiquitous presence of lipids, small amounts of cryptic triesters could easily remain undetected. On the other hand, the plant latex is a less lipophilic matrix, and deserves a systematic investigation, which could also provide interesting information on the dynamic of transport of tigliane esters from their site of production to that of their storage in the seeds.

3.2 Spectroscopic Properties and Structure Elucidation

The tigliane skeleton is based on a 5(A)/7(B)/6(C)/3(D) tetracyclic system featuring A/B-*trans*, B/C-*trans*, and C/D-*cis* junctions. Unlike the ingenane skeleton, its two-dimensional representation is straightforward, and generally drawn with the cyclopentane ring A on the left-handed lower quadrant, and the cyclopropane D ring

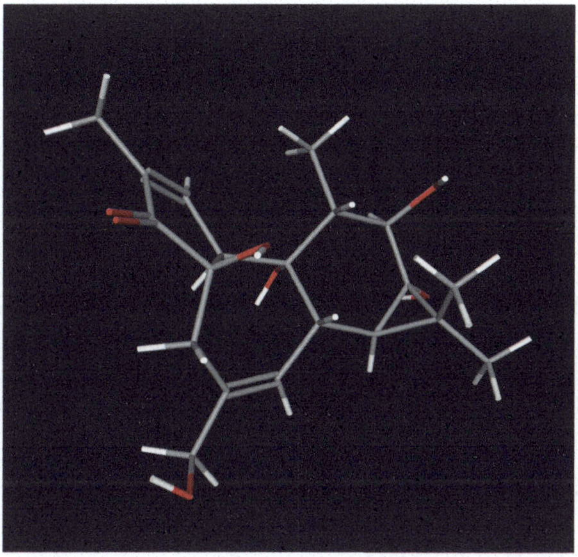

Fig. 3 Three-dimensional view of phorbol (**1**), wire frame model from PubChem at https://pubchem.ncbi.nlm.nih.gov

on the right-handed upper quadrant. Wedges and broken lines are used to indicate the orientation of the substituents around the ring system(s).

The isoprenoid skeleton of phorbol (**1**) is rigid, with little conformational mobility. Crystallographic studies on phorbol and its derivatives (see Sect. 3.2) have shown that ring A has an envelope conformation with C-1 as the flap, ring B is a distorted envelope with the fold through C-4 and C-8, and ring C adopts a half-chair conformation to accommodate the flat cyclopropane D ring (Fig. 3).

The rigidity of the framework makes steric differentiation between the α- and the β-faces excellent for addition reactions at C-12 and at the ring B double bond, but stereodifferentiation is suboptimal for addition at the C-3 carbonyl group (see Sect. 4.1). The structure of phorbol is strained at its two termini, the cyclopentane A ring and the cyclopropane D ring. The former is thermodynamically more stable when *cis*-fused to the cycloheptane B ring, while the latter is easily opened under acidic or oxidative conditions. In fact, the oxidative lability of cyclopropanol was discovered with phorbol (**1**) and later confirmed as a general feature for this class of compounds (see Sect. 4.3). As a result of such strain, conjugation in the ring A enone system is suboptimal, as evidenced by UV and NMR spectroscopy (see below).

UV and IR Spectroscopy

Phorbol (**1**) shows UV maxima in ethanol at 235 nm ($\varepsilon = 5{,}200\ M^{-1}\ cm^{-1}$) and 334 nm ($\varepsilon = 70\ M^{-1}\ cm^{-1}$). The ε value of the cyclopentenone $\pi \rightarrow \pi^*$ transition is rather low, implying a suboptimal conjugation, and this was associated with the strained *trans*-junction with ring B [52]. However, the more stable 4α-isomer (**14**) shows a similar ε value for the $\pi \rightarrow \pi^*$ transition, suggesting the involvement of additional factors for the low ε value. Since ε is increased to 6,050 M^{-1} cm^{-1} in 4-deoxyphorbol esters [53], the formation of intramolecular H-bonding between the cyclopentenone carbonyl and the 4-hydroxy group could be at the origin of the hypochromic effect, and undermine the electronic bias of the carbonyl for planarity with the Δ^1-double bond.

The IR spectrum of **1** is affected by the nature of the retained solvent. Thus, depending on the co-crystallized solvent, the (C=O) and the (C=C) IR stretching frequencies of the enone system have been reported, respectively, from 1730 to 1700 cm^{-1} and from 1640 to 1630 cm^{-1}, with the (C=O) stretching band sometimes being duplicated, as in the ethyl acetate solvate [52]. Due to the occurrence of phorbol esters in a triglyceride matrix, IR spectroscopy cannot be reliably used for the quantification of phorbol esters in the seed oils from euphorbiaceous plants [54].

NMR Spectroscopy

The ^{1}H and ^{13}C NMR spectra of tigliane esters show little overlapping between the signals of the diterpenoid protons and those of the diterpenoid carbons, simplifying the structure elucidation of this class of compounds. Figure 4 shows the assignments of ^{1}H and ^{13}C-NMR resonances for the parent polyol, phorbol (**1**), in DMSO-d_6 as the solvent [54].

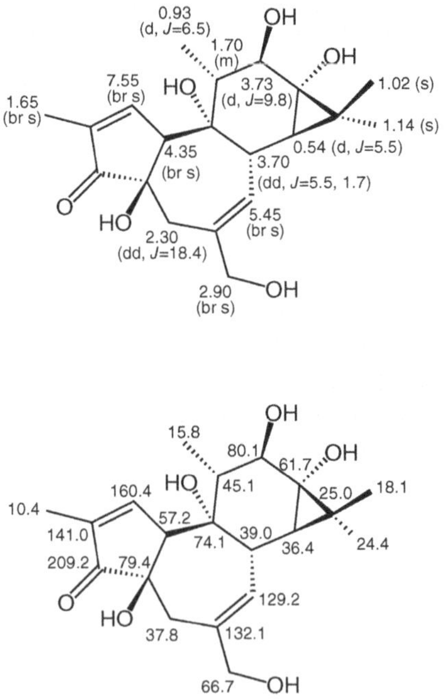

Fig. 4 Assignments of ^{1}H (top) and ^{13}C NMR (bottom) spectra of phorbol (**1**) (DMSO-d_6, 400 and 100 MHz, respectively, δ_H and δ_C in ppm)

The hydrogen resonances span the chemical shift interval between the downfield signal of H-1 ($\delta = 7.55$ ppm, s) and the upfield one of the cyclopropane C-14 proton ($\delta = 0.54$ ppm, d, $J = 5.5$ Hz). The signal of H-1 resonates in a spectroscopic area scarcely populated in lipophilic extracts and can be used to quantify the presence of tigliane esters in crude extracts and in euphorbiaceous seed oils using the H-8 signal of caffeine ($\delta = 7.56$ ppm in $CDCl_3$) as internal standard [36]. The signal of the methines H-8 and H-10 resonate rather downfield, at, respectively, around $\delta =$ 3.80 ppm (d, $J = 5.4$ Hz), and $\delta = 4.35$ ppm (br s), but no other significant chemical shift peculiarities are present in the ^1H NMR spectrum. Conversely, the ^{13}C NMR spectrum shows some unexpected features. The resonances are spread between those of the cyclopentenone carbonyl C-4 ($\delta = 209.2$ ppm, s) and the allylic methyl C-19 ($\delta = 10.4$ ppm, q). Of note, the upfield chemical shift of the tertiary cyclopropanol carbon C-13 ($\delta = 61.7$ ppm, s) is in line with the effect of hybridization on the parent system, and contrasts with the relatively downfield chemical shift of the two other cyclopropane carbons ($\delta = 36.4$ ppm, d for C-14 and $\delta = 25.0$ ppm, s for C-15). Also surprising is the observation that the secondary oxymethine C-12 resonates downfield ($\delta = 80.1$ ppm, d) of the tertiary hydroxylated carbons C-4 ($\delta = 79.4$ ppm, s) and C-9 ($\delta = 74.1$ ppm, s). The remarkable difference between the chemical shifts of the cyclopropane *gem*-dimethyls is due to γ-shielding experienced by the β-methyl group (C-17, $\delta = 18.1$ ppm, q) from the *syn*-oriented C-12 and C-8 carbon atoms. This shielding is not experienced by the C-16 methyl, which resonates at $\delta = 24.4$ ppm (q). These general features are also observed in the modified tigliane polyols 4- and 12-deoxyphorbol, 16-hydroxyphorbol, and 5-hydroxy-6,7-epoxyphorbol.

Two empirical correlations have emerged from the analysis of the ^1H NMR spectra of tigliane esters:

1. The position of H-1 in $CDCl_3$ can differentiate esters of phorbol (**1**) from those of 4α-phorbol (**14**). In compounds from the native series, H-1 resonates downfield close to the $CDCl_3$ residual proton at $\delta = 7.27$ ppm, around $\delta = 7.30$–7.40 ppm. Conversely, in compounds from the 4α-series, the signal of H-1 appears upfield of the $CDCl_3$ residual proton, around $\delta = 6.90$–7.10 ppm.
2. The diastereotopic H-20 protons are quasi-isochronous when the 20-hydroxy group is free, but form a well-defined A,B-system ($J =$ ca. 12 Hz) in the 20-esters.

The nature of the acyl decoration of the parent tigliane polyol can be established by combining information from NMR and mass spectra, and of great help is, in this context, the availability of tabulated ^1H and ^{13}C NMR values for the most common ester groups bound to isoprenoid alcohols [55]. The location of ester groups at C-12 and C-20 can be secured by HMBC measurements, since in both cases proton(s) are present on the carbon bearing the acylated hydroxy group, while for the tertiary 13-ester groups NOE-measurements with the α-oriented cyclopropane methyl H_3-16 and with H-8 are highly diagnostic. The two other tertiary hydroxy groups, located at C-4 and C-9 are never found acylated in natural tiglianes (see Sect. 3.3).

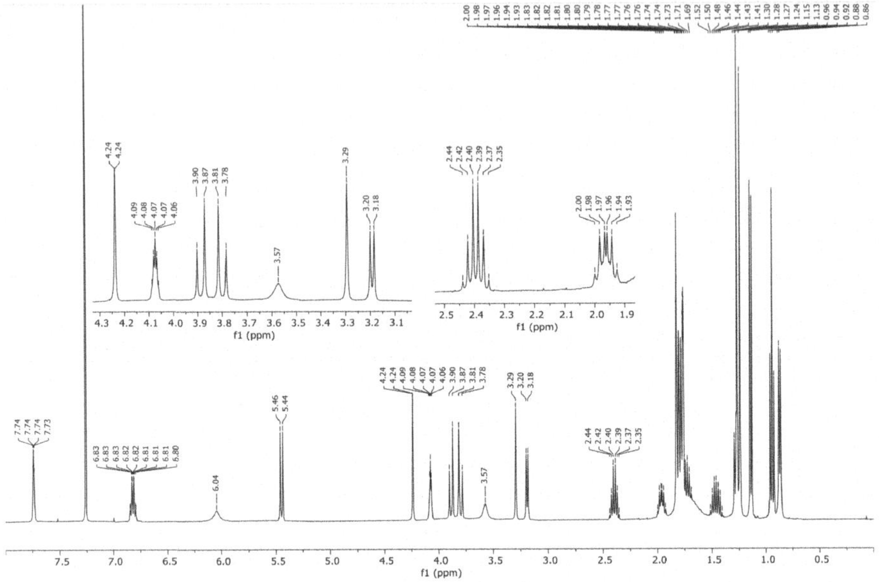

Fig. 5 ^{1}H NMR spectrum of tigilanol tiglate (**3**) ($CDCl_3$, 400 MHz, residual $CDCl_3$ proton at δ = 7.27 ppm as reference)

Acylation shifts aside, ester decoration with acids bearing aromatic ring(s) does not change the basic ^{1}H NMR spectroscopic features of the parent polyol, unlike what is observed, for instance, in certain taxane pseudoalkaloids, where esterification with acids bearing aryl groups can dramatically affect ring protons, even bringing them to negative δ-values [56]. Figure 5 shows the ^{1}H NMR spectrum of the anticancer epoxytigliane diester tigilanol tiglate (**3**), where resonances of the core tigliane skeleton are easily identified [8].

Mass Spectrometry and Mass Spectrometry-Based Hyphenated Techniques

Mass spectrometry is a powerful technique for the identification of the basic structure of tigliane esters, providing, under widely available ionization techniques (chemical- or electrospray ionization), information on the ester decoration and the nature of a tigliane polyol. In early studies based on electron-impact ionization, it was noticed that in phorbol 12,13-diesters, a long-chain ester group at C-12 is lost as an acyloxy radical (RCOO·), but as a neutral molecule (RCOOH) from C-13. Based on this observation, diesters could be sorted out into the A- and B-type, depending on the location of the long-chain acyl moiety at C-12 (A-group) or C-13 (B-group) [57]. The intensity of the molecular ion is in the range of 1% or even lower in the electron-impact ionization spectra, but chemical-ionization can provide more intense

M+1/M−1 peaks, facilitating the assignment of molecular weight. Mass spectrometry can complement NMR spectroscopy to locate functionalization in acyl moieties having functional groups (double bond(s), hydroxy, carbonyl) separated by several chain methylenes. Under electrospray ionization (ESI), double bond location can be assessed by ozone-induced dissociation (OzID), a method based on in situ ozonolysis [58], or, alternatively, by the previous photochemical Paternó-Büchi reaction with 2-acetylpyridine [59].

The coupling of LC-HRMS and MS/MS is a powerful technique to analyze complex mixtures of tigliane esters, identifying the polyol(s) present in the mixture and the molecular weight of its acyl moieties. The polyol parent peak is generally detectable, with an intensity highly variable between the various ester types, depending on the number and location of the acyl moieties. When interfaced with bioactivity data, tandem MS-based molecular networking can identify putative active molecules in bioactive extracts, accelerating dereplication and focusing the purification steps on the most active constituents of an extract [60]. For this purpose, a crude extract is fractionated, and the fractions are then evaluated by bioassay and next by MS/MS. By calculating the relative abundance of a specific compound in a fraction and comparing it to the activity level of the fraction, a bioactivity score can be calculated for a single constituent, providing a quick alternative to classical bioassay-guided fractionation for the identification of active constituents in complex extracts. Structural identification is assisted by the Global Natural Product Social (GNPS) molecular networking web platform (at http://gnps.ucsd.edu), which makes possible for automatic spectral mining in a database of MS/MS spectra of thousands of compounds. This strategy has been applied successfully to the identification of antiviral (chikungunya virus) tigliane esters in extracts of *Euphorbia dendroides* L. [60]. Remarkably, these compounds are minor constituents of the plant extract, which is made up mostly by macrocyclic jatrophane polyesters, and inactive in this assay.

Crystallographic Studies

With some notable exceptions, including phorbol myristate acetate (**2**), native tiglianes are generally amorphous solids or gums, and are not amenable to crystallographic studies. These have been reported on the unstable ethanol solvate of phorbol (**1**) [61], on its derivatives (e.g. neophorbol 13,20-diacetate-3-*p*-bromobenzoate (**8**) [62]), semi-synthetic esters (20-(5-bromofuroate) **7** [63]) or more modified derivatives, like a 20-deoxy-20-chlorotigilanol tiglate adduct with triphenylphosphine oxide (**15**) [8]. 4α-4-Deoxyphorbol-12-tigliate-13-isobutyrate (**16**), possibly an artefact from its corresponding 4β-configured native isomer, is one of the few tiglianes directly obtained from isolation solved by X-ray analysis [64].

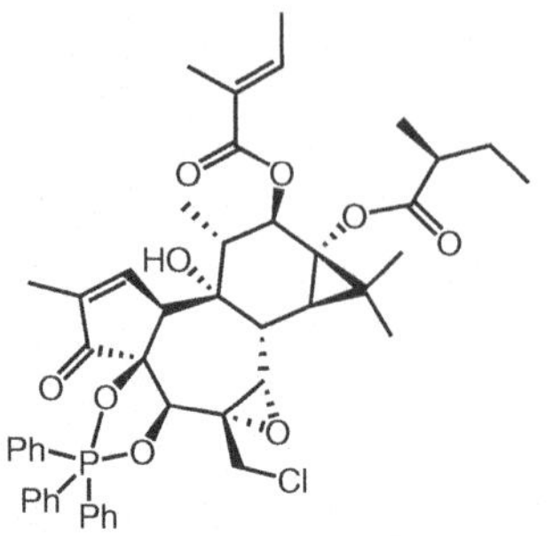

15 (20-deoxy-20-chlorotigilanol tiglate triphenylphosphine adduct)

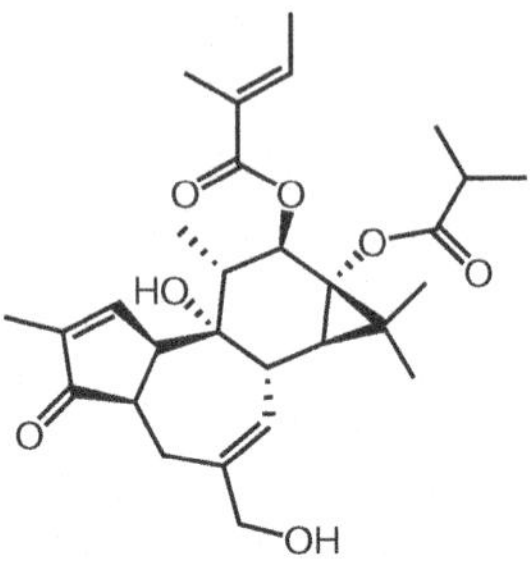

16 (4α-4-deoxyphorbol-12-tigliate-13-isobutyrate)

Optical Rotation and Electronic Circular Dichroism Studies

While the optical rotation of ingenol has caused considerable confusion in the literature [65], phorbol (**1**) has always been reported as a dextrorotatory compound, with optical rotations ranging between +100° and +120° depending on the solvent and the concentration used. Conversely, the electronic circular dichroism (ECD) of **1** originally was wrongly correlated to that of other chiral cyclopentenones, leading to the assignment of an incorrect absolute configuration of the natural product [66]. The ECD spectrum of phorbol in dioxane shows negative bands at 346 ($\Delta\varepsilon = -1.11\ M^{-1}\ cm^{-1}$) and 270 ($\Delta\varepsilon = -1.51\ M^{-1}\ cm^{-1}$), and a larger positive band at 250 ($\Delta\varepsilon = +2.8\ M^{-1}\ cm^{-1}$). Calculations with ester derivatives fit well with the observed spectra, but in protic solvents the intensity of the negative bands is decreased, while the positive band is unaffected in intensity but shifted at slightly lower wavelength values. In esters of 4α-phorbol (**14**) and its 4-deoxy derivative, the sign of the Cotton effect is reversed for both the π–π^* and the n–π^* transitions [66].

3.3 Distribution and Structural Diversity

Tigliane derivatives have a point-like distribution in Nature, being typical of plants from the Euphorbiaceae and the Thymelaeaceae families. Both families are, however, large, encompassing, respectively, 228 and 52 genera, with over 8,000 and 900 species, resulting in a large number of potential plant sources for these compounds [67]. Tigliane derivatives have a broader distribution compared to the two other classes of phorboids (ingenanes and daphnanes), in having been isolated from over thirty genera. Conversely, ingenanes only occur in three genera, and daphnanes in less than fifteen genera, with uncertainties in species calculation related to issues of botanical taxonomy [67]. The largest sources of tigliane derivatives are plants from the genera *Euphorbia* and *Croton*. While the presence of tiglianes and ingenanes is, with only a few exception(s), mutually exclusive (see Sect. 3.4), the co-occurrence of tiglianes and daphnanes is, conversely, not uncommon.

Plate 2 *Fontainea picrosperma* C.T. White. Water color painting by Ellis Rowan (1916)

Tigliane esters are lipophilic compounds, found especially in latices and in fatty seeds, where their concentration is, however, much lower than those of the dominant triterpenoids, triglycerides and fatty acids. A significant accumulation has so far been reported only in the seed kernels of *Croton tiglium* (Plate 1; see Sect. 2) and *Fontainea picrosperma* (Plate 2), where the concentration of tigliane derivatives can reach 0.3% [32] and >2% [36], with higher concentrations (ca. 1% and up to 6%) being found in the kernel seed oils. Important commercial sources of tigliane esters are the seed oils from *C. tiglium* (croton oil), *Jatropha curcas* (jatropha oil) (Plate 3), and *Sapium fordii* Hemsl. (*Vernicia fordii* (Hemsl.) Airy Shaw) (Plate 4) (tung oil). These oils are used for the production of biodiesel (croton and jatropha oils) or for the preparation of paints (tung oil). From croton oil, making up ca. 30% of *C. tiglium* seeds, the parent tigliane polyol phorbol (**1**) can be obtained in a ca. 1% yield [32].

1 (phorbol)

Fig. 6 Main modifications of the phorbol oxygenation profile occurring in natural tigliane esters

Unfortunately, oxygenation at C-16 makes the major polyols from jatropha- and tung oils unstable in the basic conditions required for their hydrolysis [38] (see Sect. 4.4). In addition, the concentration of tiglianes in tung oil is at least one order of magnitude lower than in croton oil [38].

Plate 3 *Jatropha curcas* L. Photograph courtesy R. K. Henning at www.jatropha.org; Creative Commons 2.5

Plate 4 *Vernicia fordii* (Hemsl.) Airy Shaw. Photograph courtesy KENPAI; Creative Commons 3.0

Starting from the pioneering work of Hecker [32] and Evans [33] in the 1960s and 1970s of the past century, the taxonomy of naturally occurring tigliane diterpenoids has expanded significantly, reaching an inventory of over 300 different compounds. Compared to other classes of diterpenoids, the structural diversity of tiglianes is, however, relatively limited, and focused on a combination of oxygenative/deoxygenative modifications of phorbol (**1**) with changes in the acyl decoration of the various hydroxy groups. In addition to **1**, major polyols have been described, differing from the parent compound in terms of deoxygenation at C-4, C-12, or C-20, epoxidation of the $\Delta^{6,7}$-double bond, and hydroxylation of C-5 and C-16 (Fig. 6).

Within the tigliane derivatives, some point-like mutations only occur in combination, as shown by the epoxidation of the ring B double bond, always observed as being associated with hydroxylation of C-5. Oxygenation is stereoselective, with epoxidation of the ring B double bond always occurring from the concave α-face, with hydroxylation of the C-5 methylene almost exclusively seen from the convex β-face.

We have summarized the structural diversity of tiglianes in a series of tables. Table 1 reports the ester groups found in tigliane derivatives. Some acyl groups are unique to tigliane esters, like the cyclobutane- and bicyclo[3.1.0]-dicarboxylates found in *Jatropha* factors C1–C6 (**17–21**) [68] and pycnocomolides (**22–24**) [69]. The other tables (Tables 2–26) report all known tigliane esters, along with their plant sources. Authority abbreviations for the binomials were assigned using the International Plant Name Index [67] and could differ from those used in the original reports. In the primary literature, different systematic names have been used for the same species, and their pooling in the tables under the same name has been carried out according to reference [70].

Jatropha factors C1–C6

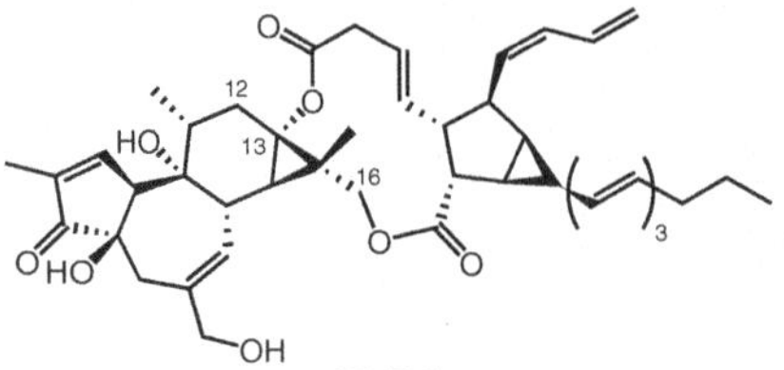

17 (C1)

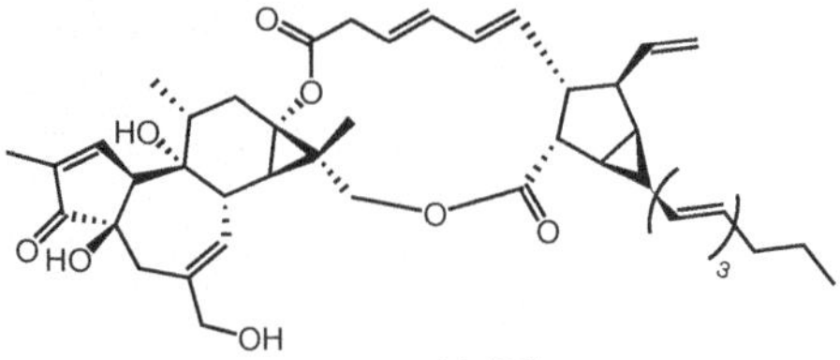

18 (C2)

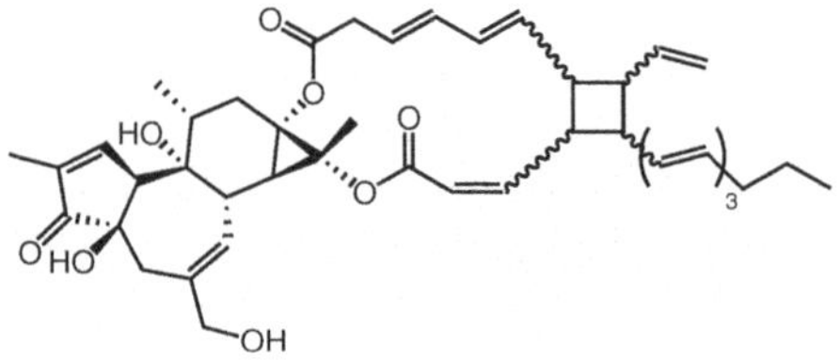

19 (C3)

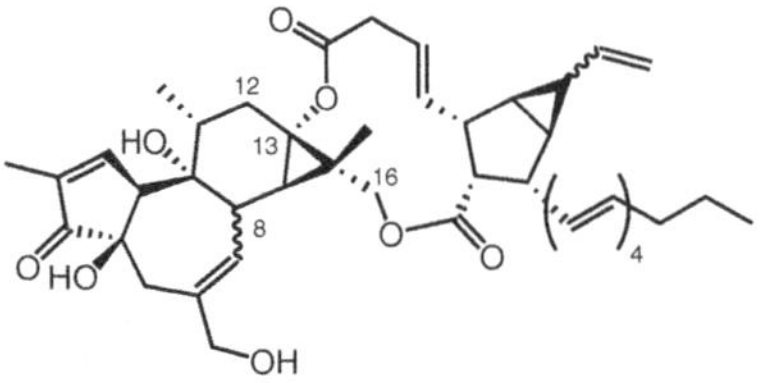

20 (C4 + C5; epimeric mixture)

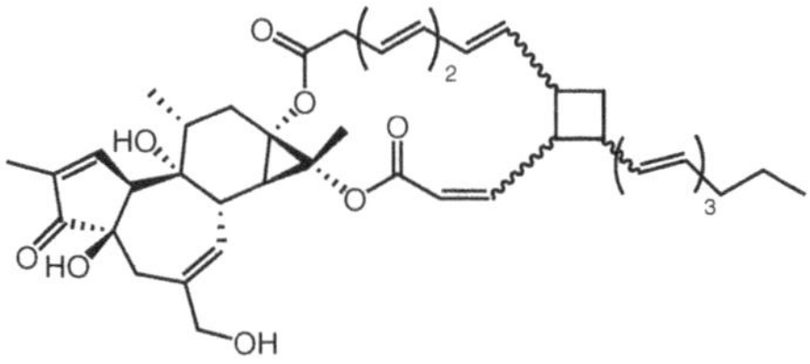

21 (C6)

Table 1 Ester groups found in phorboids

Abbreviation	Name of corresponding acid	Formula of corresponding acid
Ac	Acetic acid	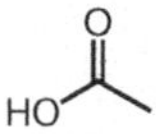
Pr	Propionic acid	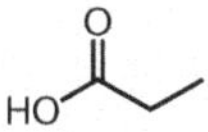
Bu	Butanoic acid	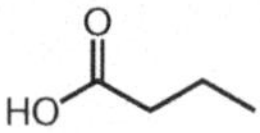
*i*Bu	Isobutanoic acid	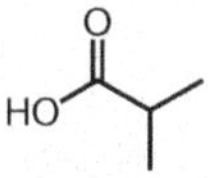
2MeBu	2-Methylbutanoic acid	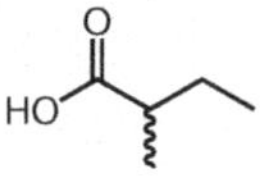
2,3MeBu	2,3-Dimethylbutanoic acid	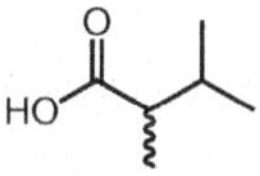
(*S*)-2MeBu	(+)-(*S*)-2-Methylbutanoic acid	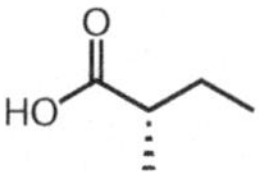
3MeBu	3-Methylbutanoic acid (isovaleric acid)	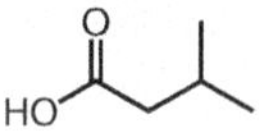
Ang	(2*E*)-2-Methyl-2-butenoic acid (angelic acid)	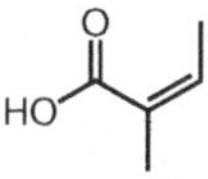
Tigl	(2*Z*)-2-Methyl-2-butenoic acid (tiglic acid)	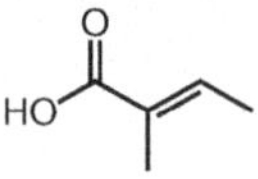
Sen	3-Methylbut-2-enoic acid (senecioic acid)	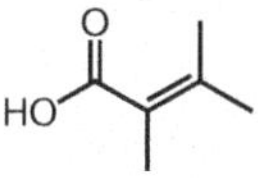
(2*S*,3*S*)-3-OH-2MeBu	(2*S*,3*S*)-3-Hydroxy-2-methylbutanoic acid	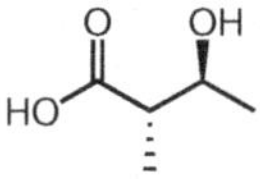
(*R*)-2-Me_2NH_2-3MeBu	(*R*)-2-(*N*,*N*-Dimethylamino)-3-methyl butanoic acid	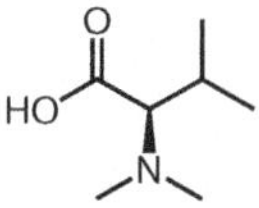

(continued)

Table 1 (continued)

Abbreviation	Name of corresponding acid	Formula of corresponding acid
(*S*)-2-Me_2NH_2-3MeBu	(*S*)-2-(*N,N*-Dimethylamino)-3-methyl butanoic acid	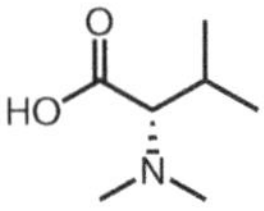
Hex	Hexanoic acid (caproic acid)	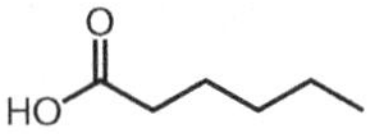
(2*E*,4*E*)-6-Oxo-hex	(2*E*,4*E*)-6-Oxohexa-2,4-dienoic acid	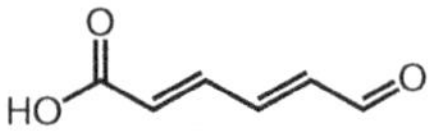
Hept	Heptanoic acid (enanthic acid)	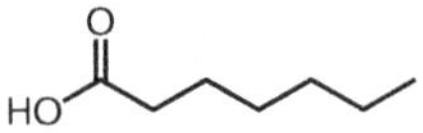
Oct	Octanoic acid (caprylic acid)	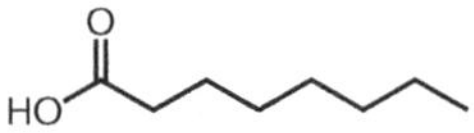
Oct:1	Octenoic acid (double bond location unspecified)	
Δ^2-Oct	2-Octenoic acid	
$\Delta^{2,4}$-Oct	2,4-Octadienoic acid	
(2*Z*,4*E*)-Oct	(2*Z*,4*E*)-Octa-2,4-dienoic acid	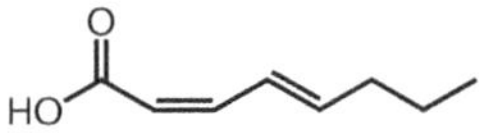
(2*E*,4*E*,6*E*)-8-Oxo-oct	(2*E*,4*E*,6*E*)-8-Oxoocta-2,4,6-trienoic acid	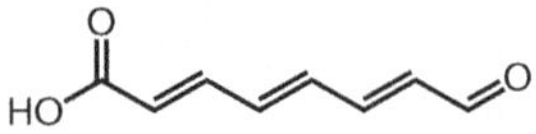
Non	Nonanoic acid (pelargonic acid)	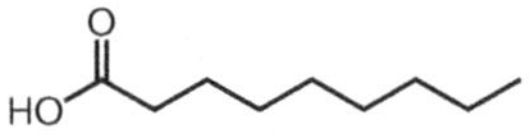
(2*Z*,4*E*,6*E*)-Non	(2*Z*,4*E*,6*E*)-Nona-2,4,6-trienoic acid	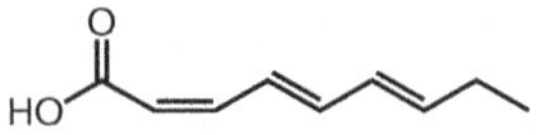
Dec	Decanoic acid (capric acid)	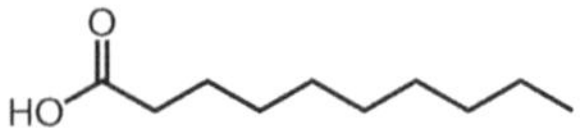
Dec:2	Decadienoic acid (double bond location unspecified)	

(continued)

Table 1 (continued)

Abbreviation	Name of corresponding acid	Formula of corresponding acid
(3*E*)-Dec	(3*E*)-Dec-3-enoic acid	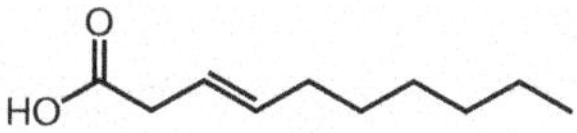
(4*Z*)-Dec	(4*Z*)-Dec-4-enoic acid (obtusilic acid)	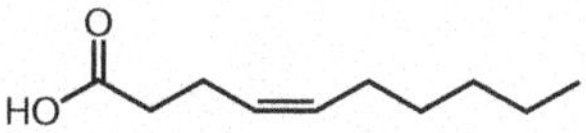
$\Delta^{2,4}$-Dec	2,4-Decadienoic acid	
(2*E*,4*E*)-Dec	(2*E*,4*E*)-Deca-2,4-dienoic acid	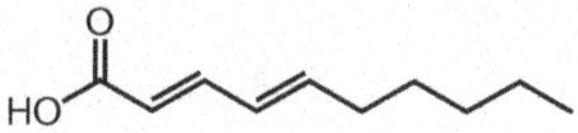
(3*E*,5*E*)-Dec	(3*E*,5*E*)-Deca-3,5-dienoic acid	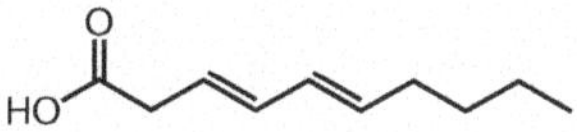
(4*Z*,7*Z*)- Dec	(4*Z*,7*Z*)-Deca-4,7-dienoic acid	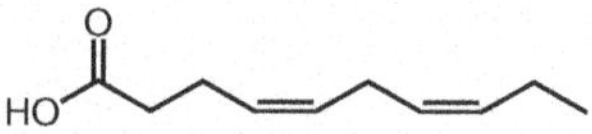
$\Delta^{2,4,6}$-Dec	2,4,6-Decatrienoic acid	
(2*Z*,4*E*)-$\Delta^{2,4,6}$-Dec	(2*Z*,4*E*)-Deca-2,4,6-trienoic acid	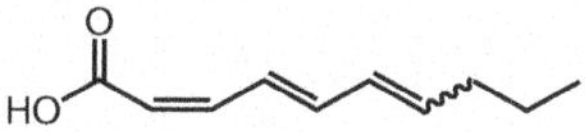
(2*E*,4*E*,6*E*)-Dec	(2*E*,4*E*,6*E*)-Deca-2,4,6-trienoic acid	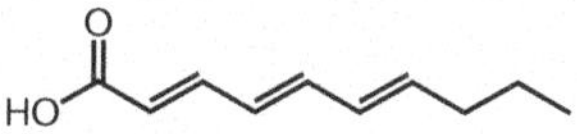
(2*E*,4*E*,6*E*,8*E*)-10-Oxo-dec	(2*E*,4*E*,6*E*,8*E*)-10-Oxodeca-2,4,6,8-tetraenoic acid	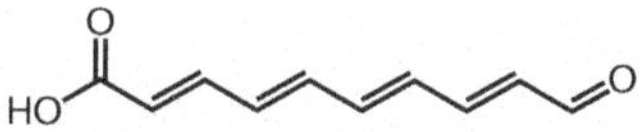
Undec:2	Undecadienoic acid (double bond location unspecified)	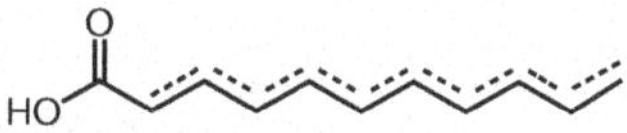
$\Delta^{2,4}$-Undec	2,4-Undecadienoic acid	
(9*S*,10*S*-Me)-Undec	(*cis*-9,10-Methylene)-undecanoic acid	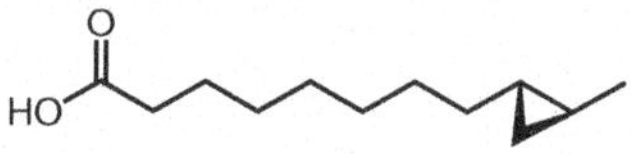
Dodec	Dodecanoic acid (lauric acid)	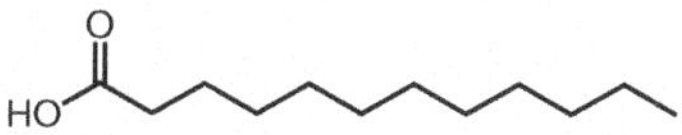
Dodec:1 Dodec:2	Dodecenoic acids (double bond locations unspecified)	

(continued)

Table 1 (continued)

Abbreviation	Name of corresponding acid	Formula of corresponding acid
$\Delta^{2,4,6}$-Dodec	2,4,6-Dodecatrienoic acid	
(2*E*,4*E*,6*E*)-Dodec	(2*E*,4*E*,6*E*)-Dodeca-2,4,6-trienoic acid	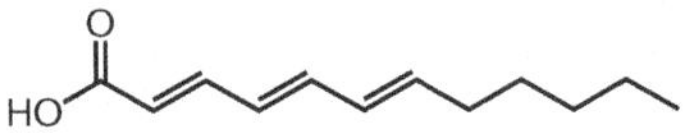
$\Delta^{2,4,6,8}$-Dodec	2,4,6,8-Dodecatetraenoic acid	
(2*E*,4*E*,6*E*,8*E*,10*E*)-12-oxo-Dodec	(2*E*,4*E*,6*E*,8*E*,10*E*)-12-Oxododeca-2,4,6,8,10-pentaenoic acid	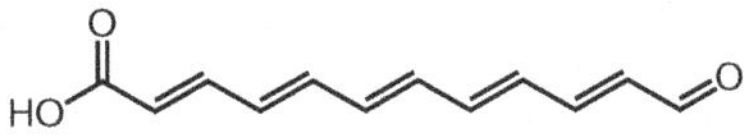
Tetradec	Tetradecanoic acid (myristic acid)	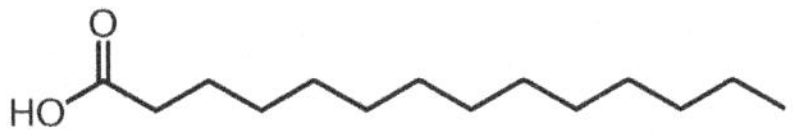
12-(Me)-Tetradec	12-Methyl-tetradecanoic acid	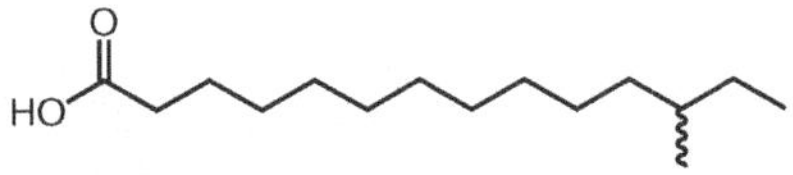
Tetradec:1 Tetradec:2 Tetradec:3	Tetradecenoic acids (double bond locations unspecified)	
(5*Z*)-Tetradec	(5*Z*)-Tetradeca-5-enoic acid	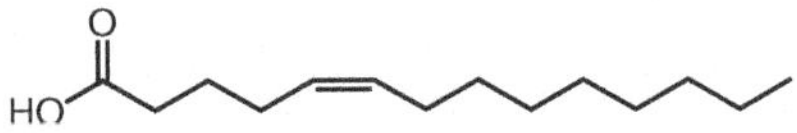
$\Delta^{2,4}$-Tetradec	2,4-Tetradecenoic acid	
(2*E*,4*E*)-Tetradec	(2*E*,4*E*)-Tetradeca-2,4-dienoic acid	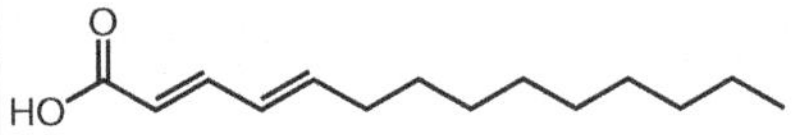
$\Delta^{3,5}$-Tetradec	3,5-Tetradecadienoic acid	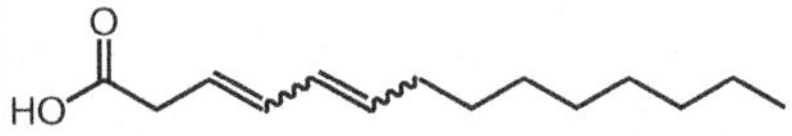
(3*E*,5*E*)-Tetradec	(3*E*,5*E*)-Tetradeca-3,5-dienoic acid	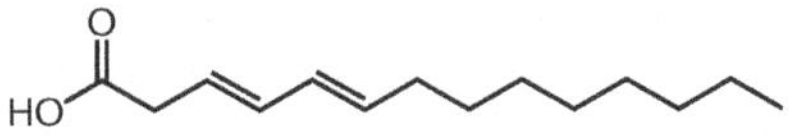
(2*E*,4*E*)-6-OH-Tetradec	(2*E*,4*E*)-6-Hydroxytetradeca-2,4-dienoic acid	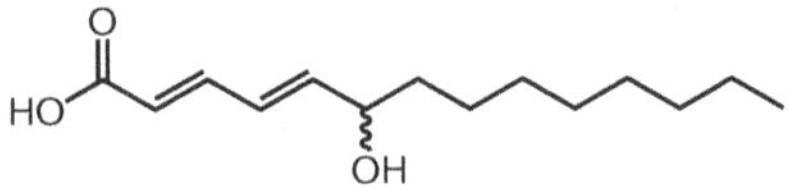
(*R*,2*E*,4*E*)-6-OH-Tetradec	(*R*,2*E*,4*E*)-6-Hydroxytetradeca-2,4-dienoic acid	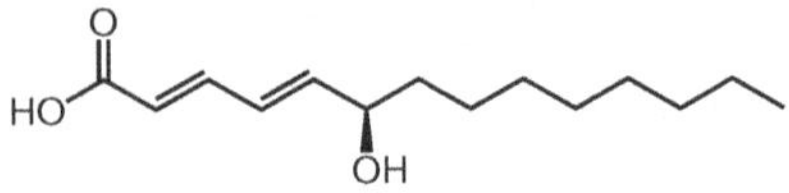

(continued)

Table 1 (continued)

Abbreviation	Name of corresponding acid	Formula of corresponding acid
(*S*,2*E*,4*E*)-6-OH-Tetradec	(*S*,2*E*,4*E*)-6-Hydroxytetradeca-2,4-dienoic acid	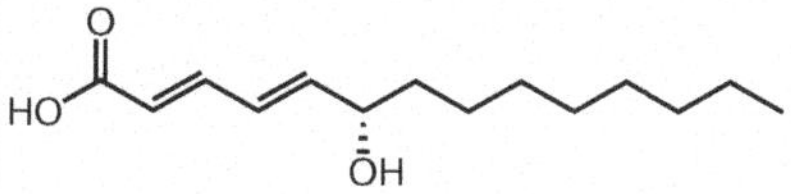
(2*E*,4*E*)-6-O-Tetradec	(2*E*,4*E*)-6-oxotetradeca-2,4-dienoic acid	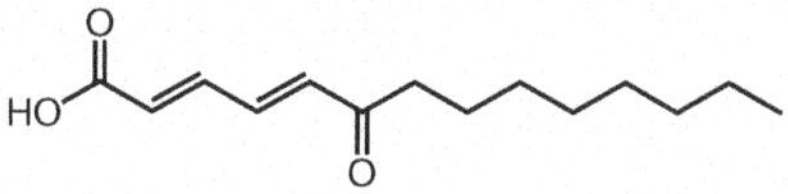
(2*E*,4*E*)-6,7-OH-Tetradec	(2*E*,4*E*)-6,7-Dihydroxytetradeca-2,4-dienoic acid (*erythro* and *threo*)	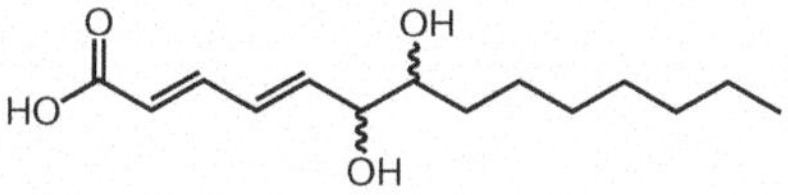
(2*Z*,4*E*,6*E*)-Tetradec	(2*Z*,4*E*,6*E*)-Tetradeca-2,4,6-trienoic acid	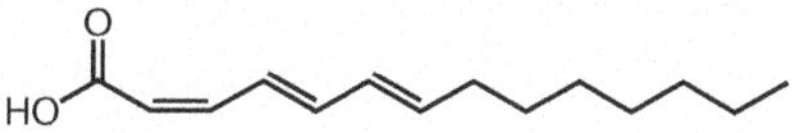
$\Delta^{2,4,6,8}$-Tetradec	2,4,6,8-Tetradecatetraenoic acid	
(2*E*,4*E*,6*E*,8*E*)-Tetradec	(2*E*,4*E*,6*E*,8*E*)-Tetradeca-2,4,6,8-tetraenoic acid	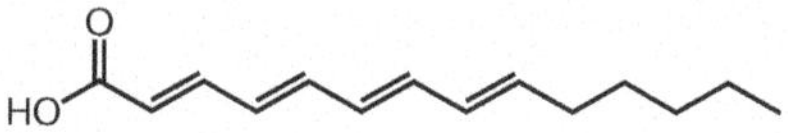
$\Delta^{2,4,6,8,10}$-Tetradec	2,4,6,8,10-Tetradecapentenoic acid	
Pentadec	Pentadecanoic acid	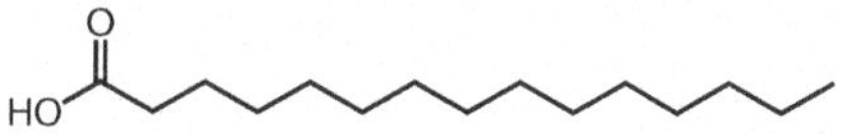
Hexadec	Hexadecanoic acid (palmitic acid)	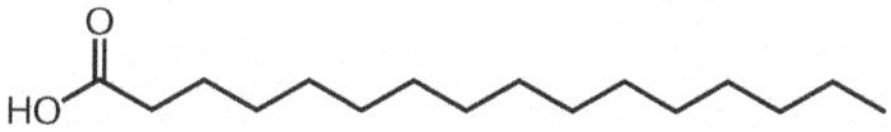
(7Z)-Hexadec	(7Z)-Hexadeca-7-enoic acid (hypogeic acid)	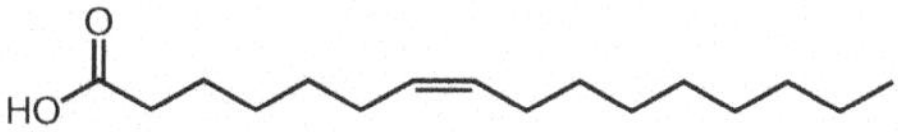
$\Delta^{3,5,7}$-Hexadec	3,5,7-Hexadecatrienoic acid	
(2*E*,4*E*,6*E*)-Hexadec	(2*E*,4*E*,6*E*)-Hexadeca-2,4,6-trienoic acid	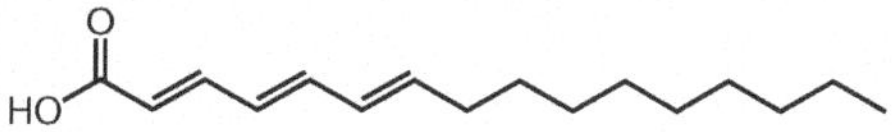
(2*E*,4*E*,6*E*)-8-Oxo-hexadec	(2*E*,4*E*,6*E*)-8-Oxohexadeca-2,4,6-trienoic acid	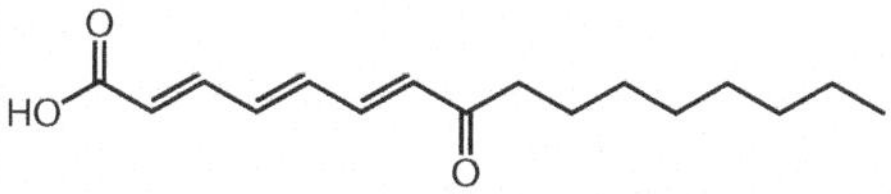

(continued)

Table 1 (continued)

Abbreviation	Name of corresponding acid	Formula of corresponding acid
Octadec:1	Octadecenoic acid (double bond location unspecified)	
(6Z)-Octadec	(6Z)-Octadec-6-enoic acid (petroselinic acid)	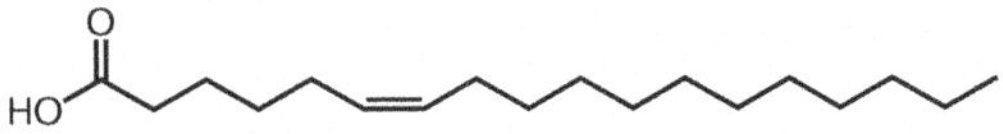
(9Z)-Octadec	(9Z)-Octadec-9-enoic acid (oleic acid)	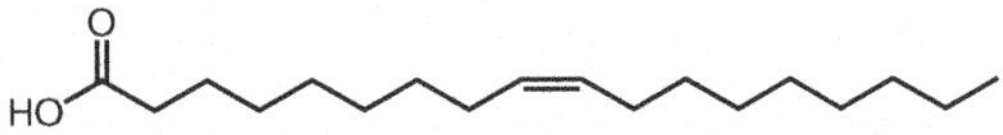
(9Z,12Z)-Octadec	(9Z,12Z)-Octadeca-9,12-dienoic acid (linoleic acid)	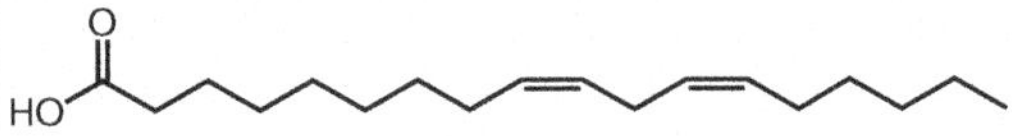
(9Z,12Z,15Z)-Octadec	(9Z,12Z,15Z)-Octadeca-9,12,15-trienoic acid (linolenic acid)	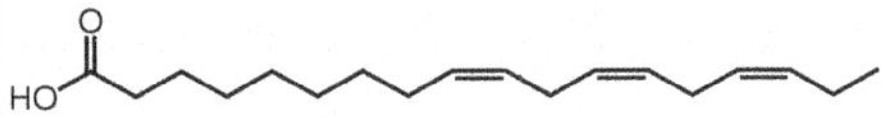
Tricosa	Tricosanoic acid	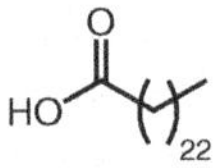
Me-2Abz-OH	*N*-Methylanthranilic acid	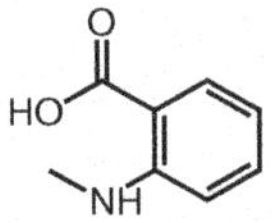
Bz	Benzoic acid	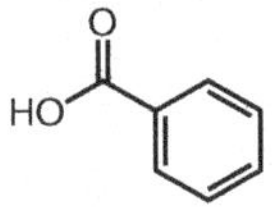
4-(OMe)Bz	4-Methoxybenzoic acid	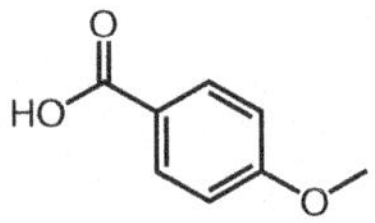
PhAc	Phenylacetic acid	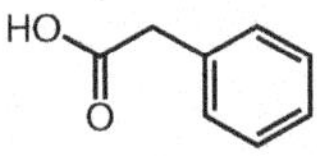
4-(OH)PhAc	4-Hydroxyphenyl acetic acid	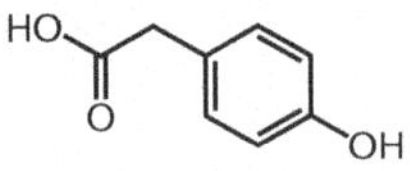
4-(OMe)PhAc	4-Methoxyphenyl acetic acid	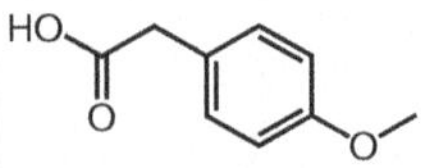
4-(OAc)PhAc	4-Acetoxyphenyl acetic acid	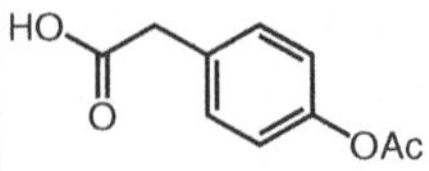

(continued)

Table 1 (continued)

Abbreviation	Name of corresponding acid	Formula of corresponding acid
—R_a—	(1*R*,2*R*,3*R*,4*S*,5*R*,6*R*)-4-((*Z*)-Buta-1,3-dien-1-yl)-3-((*E*)-3-carboxyprop-1-en-1-yl)-6-((1*E*,3*E*,5*E*)-nona-1,3,5-trien-1-yl)bicyclo[3.1.0]hexane-2-carboxylic acid	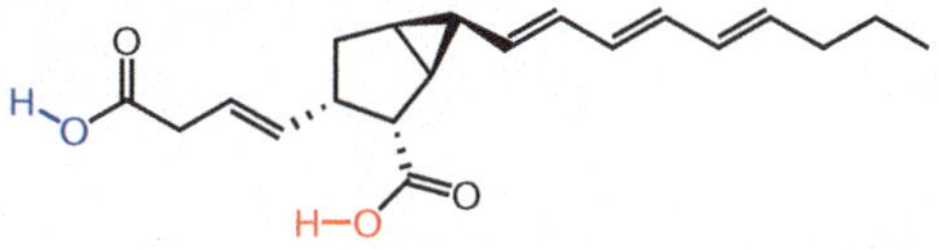
—R_b—	(1*R*,2*R*,3*R*,4*R*,5*R*,6*S*)-3-((1*E*,3*E*)-5-Carboxypenta-1,3-dien-1-yl)-6-((1*E*,3*E*,5*E*)-nona-1,3,5-trien-1-yl)-4-vinylbicyclo[3.1.0]hexane-2-carboxylic acid	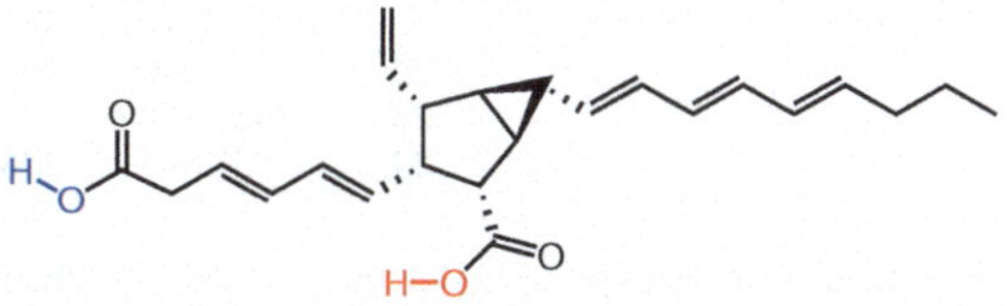
—R_c—	(1*S*,2*S*,3*R*,4*S*,5*R*,6*S*)-2-((*E*)-3-Carboxyprop-1-en-1-yl)-4-((1*E*,3*E*,5*E*,7*E*)-undeca-1,3,5,7-tetraen-1-yl)-6-vinylbicyclo[3.1.0]hexane-3-carboxylic acid	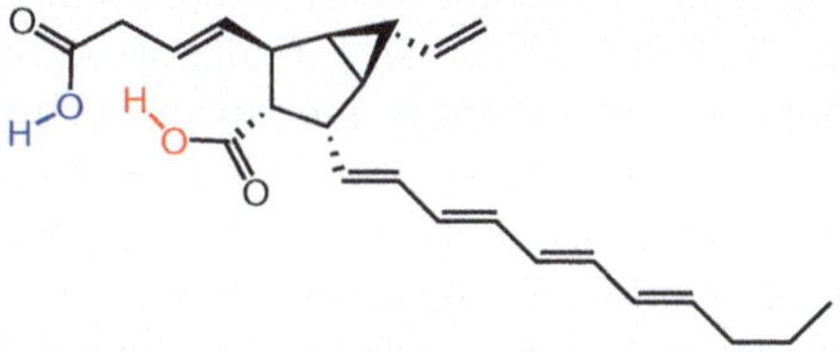
—R_d—	(1*S*,2*S*,3*R*,4*S*,5*R*,6*R*)-2-((*E*)-3-Carboxyprop-1-en-1-yl)-4-((1*E*,3*E*,5*E*,7*E*)-undeca-1,3,5,7-tetraen-1-yl)-6-vinylbicyclo[3.1.0]hexane-3-carboxylic acid	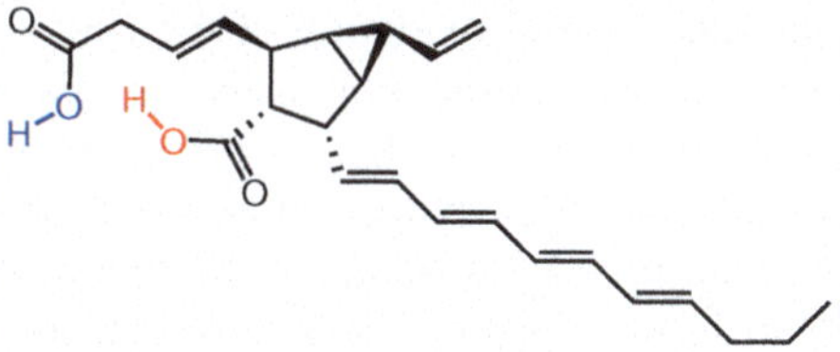
—R_e—	(3*E*,5*E*)-6-(2-((*Z*)-2-Carboxyvinyl)-3-((1*E*,3*E*,5*E*)-nona-1,3,5-trien-1-yl)-4-vinylcyclobutyl)hexa-3,5-dienoic acid	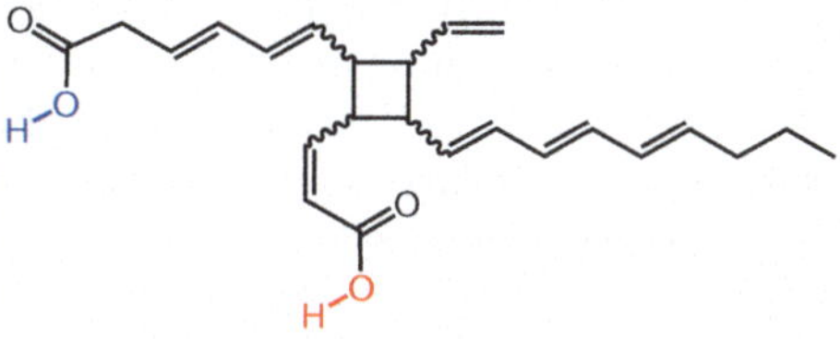

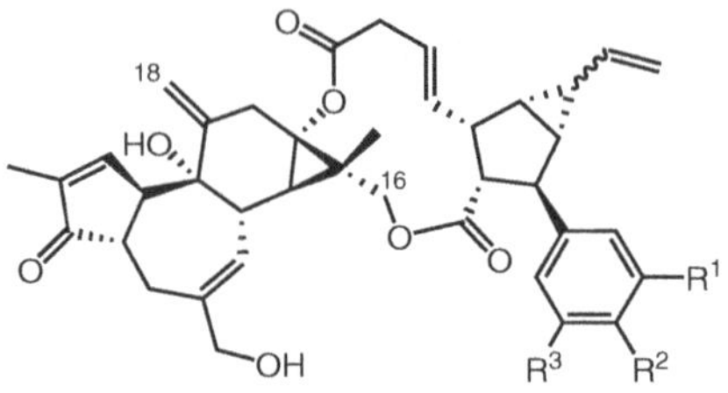

	R^1	R^2	R^3
22	H	OMe	H
23	OMe	OMe	H
24	–OCH$_2$–		OMe

For inventory purposes, we have used phorbol (**1**) as the parent polyol, prioritizing deoxygenative modifications compared to the oxygenative ones, because of their more widespread occurrence. Using deoxygenation as a "senior change", four classes of tigliane polyols can be identified, namely, phorbol esters sensu stricto, and their mono-, di-, and trideoxy derivatives, with each class also including hydroxylated versions of the basic polyol, resulting in apparently oxymoronic deoxy-hydroxy analogues of phorbol. Compounds differing from **1** because of other structural variations will be classified in the section "varia".

Given the facile aerial oxidation of the ring B double bond and the allylic 20-hydroxy group, as well as the propensity for base-induced retro-aldol epimerization at C-4 to afford tiglianes from the 4α-series (see Sect. 4), compounds showing these features are most probably extraction/isolation artefacts, but, nevertheless have been indexed, making up a large share of the inventory of tigliane diterpenoids. Thus, out of the 90 tigliane esters characterized from *Croton tiglium*, 26 are potential isolation artefacts [71]. This abundance of potential artefacts is no doubt due to the complexity of the tigliane mixtures, requiring lengthy fractionation protocols and prolonged contact with the air and chromatographic stationary phases. A significant portion of the reported tigliane diversity is therefore programmed by their reactivity rather than generated enzymatically.

Esterification Diversity

The most common esterification profile of tigliane polyols involves the hydroxy groups at C-12 and C-13. The ester groups are generally distinct, with only very few exceptions (three 12,13-bis-2-methylbutyrates, and three 12,13-bis-isobutyrates, Tables 2–26). Within the tigliane 12,13-diesters, three major acylation patterns can be distinguished, namely, that with a long-chain acyl moiety at C-12 and a short one at C-13 (type-A diesters), that with an opposite pattern (short chain, often branched, at C-12 and a long-chain at C-13, type-B diesters), and those having two short chains, often one or both branched (type-C diesters). The three series are exemplified by phorbol 12-myristate-13-acetate (**2**), phorbol-12-acetate-13-laurate (**25**) and phorbol 12-tiglate-13-butyrate (**26**), all occurring in croton oil [71]. When lists of co-occurring saturated and unsaturated diesters are compared within a specific producer, a definite strategy of acylation is evident, with unsaturation more frequent on 12-ester groups compared to the 13-acyl moieties. Thus, esterification with tiglic acid is more common at the C-12-hydroxy group, while its saturated analogue (α-methylbutyric acid) is more commonly found at the adjacent C-13-hydroxy group. Conversely, long-chain fatty acids like oleic and linoleic acid are typical of the 20-hydroxy group and relatively rare at other locations. These observations suggest the presence of specific acylases, selective for certain hydroxy groups, and for which their raison d'être remains, however, unknown.

The acyl moieties found in tigliane polyols and reported in Table 1 have been associated with a specific abbreviation. There is no consistency in the use of many of these abbreviations in the literature, and we have selected the most widespread or those that appear more logical and seem to convey more structural information.

Unsaturation and poly-unsaturation are not uncommon, but in many cases the location and configuration of the double bonds in the acyl moiety have not been defined. Terminal (ω) location of double bonds is documented, but is less common compared to non-terminal ones, as is the presence of heteroatoms, not only hydroxy and carbonyl functions, but also amino groups, with a remarkable example being in the acyl decoration of both enantiomers of *N*,*N*-dimethylvaline.

Aromatic acids and aryl-containing acids (e.g. phenylacetic acid) are more common at the 12-hydroxy group, and the occurrence of a *N*-methylanthranoyl group is interesting because of its fluorescence. Thus, the 12-*N*-methylanthranilate sapintoxin A (**27**) [72] has been developed as a fluorescence probe for membrane studies related to protein kinase C activation [73]. The linear acyl moieties are presumably of polyketide origin, with ω-2 cyclopropanation and ω-3 methylation deriving from ω-2 olefin precursors via *S*-adenosylmethionine-mediated processes [74]. It is unclear to what extent the presence of oxygenated functions (hydroxy groups and carbonyls) is associated with auto-oxidation or, conversely, to enzymatic oxidizing activity.

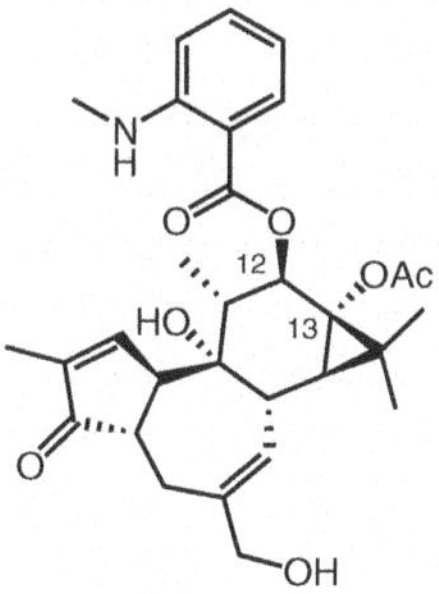

27 (12-*N*-methylanthranilate sapintoxin A)

Jatropha factors (**17–21**) [68] and pycnocomolides (**22–24**) [69] contain complex dicarboxylic acids at the C-13 and C-16 hydroxy groups of their tigliane core. Their biosynthesis could be associated with the photochemical dimerization of linear polyunsaturated esters, as depicted in Scheme 1 for the *Jatropha* factors. For the sake of simplicity, the reactions are presented as occurring with the free acids **28** and **29**, but the tigliane template undoubtedly facilitates their combination, securing the close spatial relationship referred to by Woodward when he claimed that "vicinity favors intimacy". The cyclobutane dicarboxylates **30** and **31** occur in the *Jatropha* factors C-3 (**19**) and C-6 (**21**), and presumably are the result of regiochemically distinct photochemical [2+2] cycloadditions between the polyunsaturated acids with ten (**28**) and fourteen (**29**) carbons, located, respectively, at C-13 and C-16 of the tigliane core. The C-16 polyunsaturated acid **28** reacts with its carbonyl-conjugated Δ^4-double bond, and that at C-13 (**29**) with either its Δ^7- or Δ^9- unconjugated double bond, generating, respectively, the dicarboxylated **30** and **31**, as depicted in Scheme 1 for the formation of **30**.

The bicyclo[3.1.0]hexane dicarboxylates occurring in the *Jatropha* factors C1 (**17**), C2 (**18**), and C4+C5 (**20**), could originate from three divergent [4+2] cycloadditions involving the polyunsaturated acids **28** and **29** and the formation of isomeric cyclohexene derivatives. A cycloaddition where the C-10 carbon-unsaturated acid **28** acts as the 4π-component and the fourteen carbon acid **29** as the 2π component could lead to the cyclohexene acids **32** en route to the *Jatropha* factors C4+C5 (**20**), as outlined in Scheme 1. Conversely, the diacid **33**, en route to the *Jatropha* factor C1 (**17**), may originate from an alternative [4+2] cycloaddition where the $\Delta^{2,4}$ diene system of the fourteen carbon acid **29** reacts with the Δ^5 double bond of the C-10 carboxylic acid **28** (Scheme 1), and a similar cycloaddition with the Δ^7 double bond of **28** as the 2π component would generate the cyclohexene precursor of the *Jatropha* factor C_2. A photochemical di-π-methane-like rearrangement, represented in Scheme 1 as a concerted process but most likely of the radical type, could eventually turn the cyclohexene adducts into the bicyclic dicarboxylates occurring in the *Jatropha* factors, as outlined in Scheme 1 for the conversion of **32** (or otherwise, **33**) into the dicarboxylic acid **34** occurring in the *Jatropha* factors C4+C5 (**20**) (obtained as a non-resolved epimeric mixture). A mechanistically similar process where the C-14 acid is replaced by the cinnamic acid derivatives (**35–37**) could generate the dicarboxylic acids of pycnocomolides A–C (**22–24**).

The acyl moieties bound to tigliane polyols are generally lipophilic, but the presence of hydroxy and amino groups along the chain has also been documented (see Table 1; Sect. 3.3). This could have marked effects on bioactivity, perturbing the lipophilic interactions that have proved so important for the interaction of these compounds with macromolecular end points (see Sect. 6).

Glycosidation Diversity

Glycosidation is very rare among tigliane esters and has so far been reported in a single species of *Euphorbia* (*E. fischeriana* Steud.) [75]. The sugars involved are D-glucose and its 3-, 4-, and 6-galloyl esters. Due to their polarity, it seems possible that the occurrence of tigliane glycoconjugates is underestimated, since they would not partition well in the lipophilic solvents used for biomass extraction.

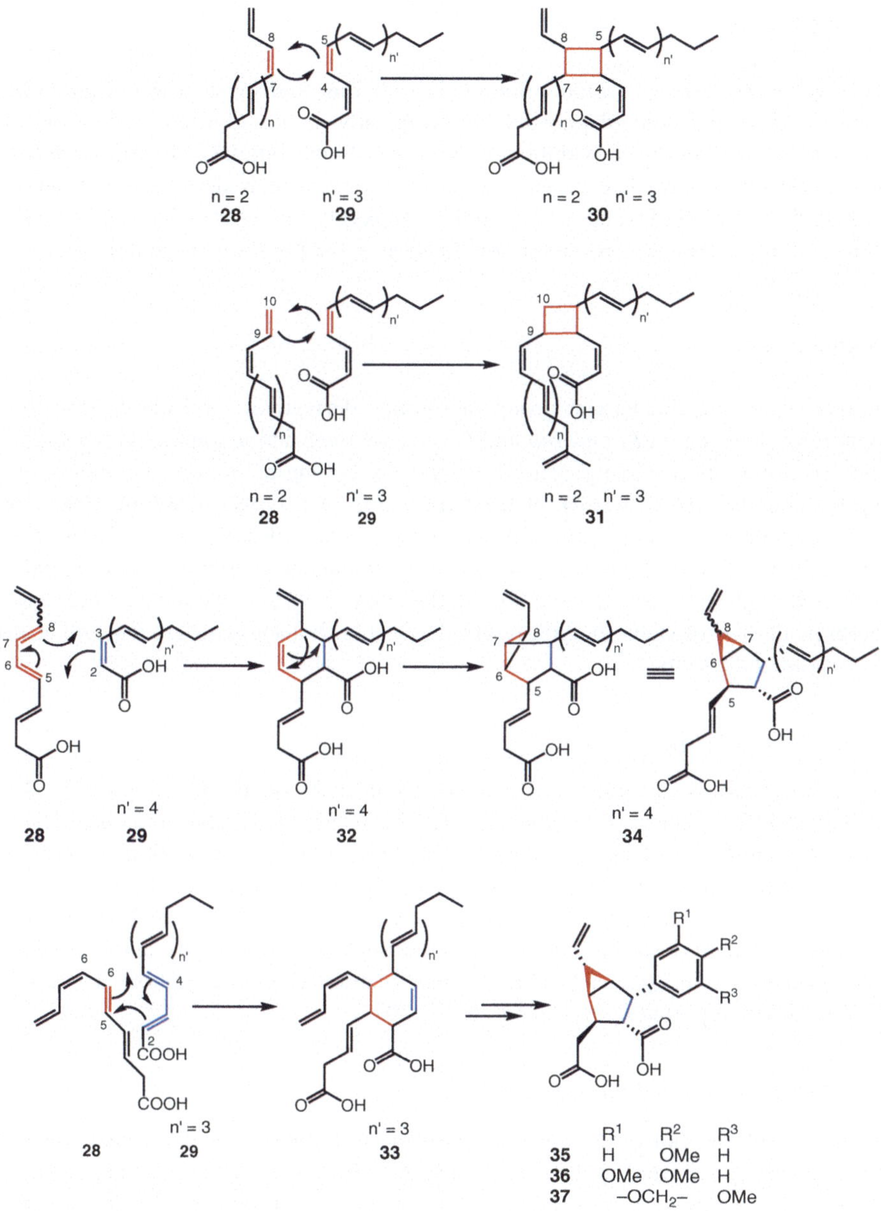

Scheme 1 Possible biogenesis of the complex dicarboxylic acids found in the *Jatropha* factors

Polyol Diversity

The skeletal diversity of tigliane polyols is very low, and most of them retain the connectivity of the pentaol phorbol, differing only in the addition or deletion of hydroxy group(s), or the epoxidation of the ring B double bond. For listing purposes, the natural esters have been sorted so as to prioritize their deoxygenated nature compared to phorbol (**1**) (and the 12-, and 20-deoxyphorbol esters), including within each class any additional oxidative modification of the phorbol oxygenation.

Phorbol Esters

Phorbol esters are the largest group of tigliane derivatives, and plants from the euphorbiaceous genera *Euphorbia* and *Croton* are their major source, with a significant contribution occurring also from species belonging to the Thymeleaceae family (*Daphne*, *Wikstroemia*), at least for esters of the parent polyol. Esters of 12-deoxyphorbol are more common than those of phorbol in the genus *Euphorbia*, but the distribution of the parent polyol is more botanically globalized and includes many non-euphorbiaceous genera. Under the entry "phorbol" we have included the parent penta-hydroxylated polyol, referred to as phorbol sensu stricto, as well as its hydroxylated versions.

Phorbol Esters Sensu Stricto

These diterpenoids generally occur as 12,13-diesters, or, more rarely, as 12,20-diesters and 12-monoesters. Very few triesters have been characterized owing to their poor extractability from the lipid matrices [36], but progress in separation technology is expected to significantly increase their inventory significantly in the future. Various types of diesters (A, B, C) can co-occur in the same plant, possibly with distinct tissue distribution, but this issue has not yet been investigated systematically, and its ecological rationale remains unclear. Table 2 lists all naturally occurring phorbol esters **2**, **25**, **26**, **38–109** reported up to December 31, 2023.

5- and 16-Hydroxylated Phorbol Esters

Hydroxylated phorbol esters are rare, with the introduction of additional oxygen functions being reported at the C-5 methylene and at one of the geminal methyls bound to C-15 (Table 3; **110–124**). Confusion may exist as to the location of the hydroxy group on the germinal methyls. The 16-methyl is α-oriented, and the 17-methyl β-oriented, and phorbol esters oxygenated at one of the germinal methyls have been generally reported as 16-esters. However, confusion exists because of both numbering and the assignment. Thus, in a study on the leaves of the tung oil tree (*Vernicia* (*Aleurites*) *fordii* Hemsl., Plate 4) a reversed numbering was used for compound **110**, and the β-methyl was referred to as the 16-methyl [113]. However, the detection of a ROESY cross-peak between H-8 and the oxymethylene suggests

Table 2 Naturally occurring phorbol esters

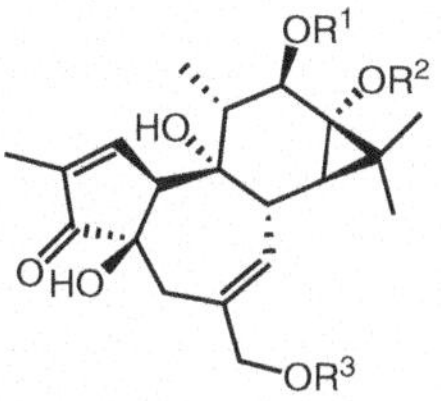

Compound	R1	R2	R3	Source	References
2	Tetradec	Ac	H	*C. tiglium* *E. stracheyi*	[32, 87, 89, 90] [76, 78]
25	Bu	Dodec	H	*C. tiglium*	[71]
26	Tigl	Bu	H	*C. tiglium*	[32]
38	H	Ac	H	*E. stracheyi* *D. aurantiaca* *E. fischeriana* *E. ebracteolata* *A. sinensis* *E. kansuensis*	[76–78] [79] [80] [81] [82] [83]
39	H	2MeBu	H	*C. tiglium*	[84]
40	H	Bz	H	*E. acerifolia*	[85]
41	Tetradec	H	H	*E. stracheyi*	[76, 78]
42	Ac	(*S*)-2MeBu	H	*Stil. lineata*	[86]
43	Ac	Tigl	H	*C. tiglium*	[87]
44	Ac	Dec	H	*C. tiglium*	[5, 32, 87–90]
45	Ac	Dodec	H	*C. tiglium*	[5, 32, 88, 89]
46	Ac	$\Delta^{2,4,6,8}$-Tetradec	H	*E. tirucalli*	[91]
47	*i*Bu	Ac	H	*C. tiglium*	[92]
48	2MeBu	Ac	H	*C. tiglium*	[92]
49	2MeBu	*i*Bu	H	*C. tiglium*	[89, 92]
50	2MeBu	2MeBu	H	*C. tiglium*	[92]
51	2MeBu	Ang	H	*E. franckiana*	[93]
52	2MeBu	Tigl	H	*C. tiglium*	[89]
53	2MeBu	Dec	H	*C. tiglium*	[87]
54	2MeBu	Dodec	H	*C. tiglium*	[89]
55	(*S*)-2MeBu	Oct	H	*C. tiglium*	[32]
56	(*S*)-2MeBu	Dec	H	*C. tiglium*	[32, 90]
57	(*S*)-2MeBu	Dodec	H	*C. tiglium*	[32]
58	Tigl	Ac	H	*C. tiglium*	[90, 92]
59	Tigl	Pr	H	*C. tiglium*	[82, 99]
60	Tigl	*i*Bu	H	*C. tiglium*	[89, 90, 92]
61	Tigl	2MeBu	H	*C. tiglium*	[87, 89, 92]
62	Tigl	Oct	H	*C. tiglium*	[32]
63	Tigl	Dec	H	*C. tiglium*	[32]
64	Tigl	Dodec	H	*C. tiglium*	[32]
65	(*R*)-2-Me_2NH_2-3MeBu	Ac	H	*C. ciliatoglandulifer*	[94]

(continued)

Table 2 (continued)

Compound	R^1	R^2	R^3	Source	References
66	(2*E*,4*E*)-6-Oxo-hex	Ac	H	*A. malaccensis*	[95]
67	Bz	*i*Bu	H	*C. tiglium*	[92]
68	Bz	Hept	H	*St. chamaejasme*	[96]
69	Bz	Oct	H	*D. aurantiaca* *W. scytophylla* *St. chamaejasme* *W. lamatsoensis*	[79] [97] [96, 98] [99]
70	Bz	Non	H	*D. aurantiaca*	[100]
71	Bz	Dec	H	*St. chamaejasme* *W. canescens*	[96, 98] [101]
72	Bz	(3*E*)-Dec	H	*W. lichiangensis*	[102]
73	Bz	(4*Z*)-Dec	H	*St. chamaejasme* *W. retusa*	[96, 98] [96]
74	Bz	(3*E*,5*E*)-Dec	H	*W. canescens* *W. lichiangensis*	[101] [102]
75	Bz	(4*Z*,7*Z*)-Dec	H	*St. chamaejasme*	[96, 98]
76	Oct	Ac	H	*P. elongata*	[103]
77	(2*Z*,4*E*)-Oct	Ac	H	*E. tirucalli*	[91]
78	(2*E*,4*E*,6*E*)-8-Oxo-oct	Ac	H	*E. kansuensis*	[83]
79	Dec	Ac	H	*D. genkwa* *C. mauritianus* *C. tiglium* *P. elongata*	[104, 105] [106] [5, 32, 88, 90] [103]
80	Dec	2MeBu	H	*C. tiglium*	[87]
81	(4*Z*)-Dec	Ac	H	*P. elongata*	[103]
82	(2*E*,4*E*)-Dec	Ac	H	*D. genkwa* *E. stracheyi*	[104, 105] [76, 78]
83	(4*Z*,7*Z*)-Dec	Ac	H	*P. elongata*	[103]
84	$\Delta^{2,4,6}$-Dec	Ac	H	*A. malaccensis* *D. genkwa*	[107] [108]
85	(2*Z*,4*E*)-$\Delta^{2,4,6}$-Dec	Ac	H	*E. tirucalli*	[91]
86	(2*E*,4*E*,6*E*)-Dec	Ac	H	*S. japonicum*	[109]
87	(2*E*,4*E*,6*E*,8*E*)-10-Oxo-dec	Ac	H	*E. kansuensis*	[83]
88	Undec:2	Ac	H	*O. paniculata*	[110]
89	Dodec	Ac	H	*C. tiglium* *C. sparsiflorus*	[5, 32, 88, 90] [111]
90	(2*E*,4*E*,6*E*,8*E*,10*E*)-12-Oxo-dodec	Ac	H	*E. kansuensis*	[83]
91	(2*Z*,4*E*,6*E*)-Tetradec	Ac	H	*A. malaccensis*	[112]
92	(2*E*,4*E*)-6,7-OH-Tetradec (*erythro*)	Ac	H	*A. malaccensis*	[95]

(continued)

Table 2 (continued)

Compound	R^1	R^2	R^3	Source	References
93	(2*E*,4*E*)-6,7-OH-Tetradec (*threo*)	Ac	H	*A. malaccensis*	[95]
94	$\Delta^{2,4,6,8}$-Tetradec	Ac	H	*E. tirucalli*	[91]
95	$\Delta^{2,4,6,8,10}$-Tetradec	Ac	H	*E. tirucalli*	[91]
96	Hexadec	Ac	H	*Al. fordii* *S. sebiferum* *C. tiglium*	[113, 114] [109] [32, 89]
97	Octadec:1	Ac	H	*S. sebiferum*	[109]
98	Me-2Abz-OH	Ac	H	*S. indicum*	[109, 115]
99	H	Ac	*β*-D-Glu	*E. stracheyi* *E. fischeriana*	[76, 78] [116]
100	H	Ac	(9Z)-Octadec	*C. tiglium*	[89]
101	H	Ac	(9Z,12Z)-Octadec	*C. tiglium*	[87, 89]
102	H	Tigl	(9Z,12Z)-Octadec	*C. tiglium*	[87]
103	Tricosa	H	Ac	*O. katharinae*	[117]
104	Ac	Ac	Ac	*C. tiglium* *Sy. grantii*	[90] [118]
105	*i*Bu	Ac	Ang	*E. franckiana* *E. coerulescens*	[93] [83, 119]
106	Bz	Oct	Hexadec	*W. scytophylla*	[97]
107	(2Z,4*E*)-Oct	Ac	Ac	*E. broteri*	[120]
108	Dodec	Ac	(9Z,12Z,15Z)-Octadec	*C. sparsiflorus*	[111]
109	Tetradec	Ac	Ac	*E. stracheyi* *C. tiglium*	[76, 78] [90]

A. = Aquilaria; Al. = Aleurites; D. = Daphne; E. = Euphorbia; C. = Croton; O. = Ostodes; P. = Pimelea; Sa. = Sapium; St. = Stellera; Stil. = Stillingia; Sy. = Synademium; W. = Wikstroemia

indeed oxygenation at the *β*-oriented methyl and not at the *α*-one, as reported in other publications on the same compound [90, 114]. If the spectroscopic analysis is correct, compound **110** and its analogues from *A. fordii* should therefore constitute a 17- and not a 16-ester, raising the issue of how correct the "old" data on these compounds are. 16-Acyloxyphorbol esters could be the precursors of daphnane esters (see Sect. 4), and a clarification on the structure of the naturally occurring tigliane esters hydroxylated at one of the geminal methyls should therefore be of significant biogenetic relevance. Their parent polyol is unstable in basic media, and has not yet been characterized.

5-Hydroxy-6,7-α-Epoxyphorbol Esters

Most tiglianes of this series have been isolated from a single species (blushberry, *Fontainea picrosperma*, the Australian rainforest plant from which tigilanol tiglate (**3**)

Table 3 Naturally occurring 5- and 16-hydroxylated phorbol esters

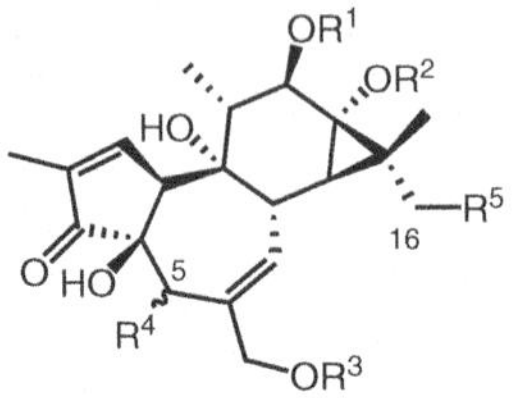

Compound	R^1	R^2	R^3	R^4	R^5	Source	References
110*	Hexadec	Ac	H	H	OH	*Al. fordii*	[90, 113, 114]
111	H	H	H	H	OH	*E. stracheyi*	[78]
112	H	H	H	H	OAc	*Sa. insigne*	[122]
113	H	Ac	H	H	OH	*Al. fordii*	[121]
114	Tigl	(*S*)-2MeBu	H	β-OH	H	*F. picrosperma*	[8]
115	Bz	Oct	H	β-OH	H	*W. lamatsoensis* *St. chamaejasme* *W. scytophylla*	[99] [96, 98] [123]
116	Bz	Dec	H	β-OH	H	*W. lichiangensis*	[102]
117	Bz	(3*E*)-Dec	H	β-OH	H	*W. lichiangensis*	[102]
118	Bz	(3*E*,5*E*)-Dec	H	β-OH	H	*W. lichiangensis*	[102]
119	Tetradec	Ac	H	H	OH	*C. flavens*	[124]
120	Hexadec	Ac	H	H	OH	*Al. fordii* *Al. fordii* *C. flavens*	[114] [113, 121] [124]
121	2MeBu	Hex	H	H	O(2MeBu)	*C. caudatus*	[125]
122	Tetradec	Ac	Dec	H	OH	*C. flavens*	[124]
123	Hexadec	Ac	Dec	H	OH	*C. flavens*	[124]
124	2MeBu	Hex	Dec	H	O(2MeBu)	*C. caudatus*	[125]

Al.= Aleurites; E. = Euphorbia; C. = Croton; Sa. = Sapium; St. = Stelleria; W. = Wikstroemia
*Hydroxylation possibly at C-17 and not at C-16

was isolated) [8, 36], and the 5β-hydroxy-α-epoxy functionalization pattern of ring B is indeed very rare in piglianes, despite its common occurrence in daphnane-type phorboids. Remarkably, epoxidation of the ring B double bond is strictly associated with the hydroxylation of the C-5 methylene, while the latter is also documented in the absence of the epoxide group (Table 3). In this context, compound **130** represents an apparent apex, in bearing a ring B-epoxidized double bond and lacking a 5-hydroxy group. However, inspection of the reported spectroscopic data [126] suggests that the epoxide is β- and not α-oriented, while the configuration of the epoxide is ambiguously referred to in the text as β- but indicated as α- in the original formula. Given the facility of auto-oxidation of phorbol esters, **130**, as suggested by the authors [126], might actually be an artefact of isolation and not a natural product.

The association of 5β-hydroxylation and α-epoxidation is surprising mechanistically, since in cyclic allylic alcohols the hydroxy group exerts a *syn*-orienting effect on epoxidation, suggesting that in the biosynthesis of these compounds epoxidation could precede hydroxylation (see also Sect. 4.5). Regarding the acylation pattern, all

Table 4 Naturally occurring 5-hydroxy-6,7-α-epoxyphorbol esters

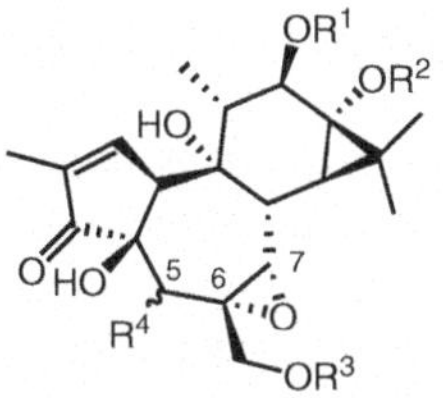

Compound	R^1	R^2	R^3	R^4	Source	References
3	Tigl	(*S*)-2MeBu	H	OH	*F. picrosperma*	[8]
125	H	(*S*)-2MeBu	H	OH	*F. picrosperma*	[8]
126	Ac	(*S*)-2MeBu	H	OH	*F. picrosperma*	[8]
127	Pr	(*S*)-2MeBu	H	OH	*F. picrosperma*	[8]
128	Bu	(*S*)-2MeBu	H	OH	*F. picrosperma*	[8]
129	(*S*)-2MeBu	(*S*)-2MeBu	H	OH	*F. picrosperma*	[8]
130*	Hexadec	Ac	H	H	*Sa. sebiferum*	[126]
131	Sen	(*S*)-2MeBu	H	OH	*F. picrosperma*	[8]
132	Hex	(*S*)-2MeBu	H	OH	*F. picrosperma*	[8]
133	(2*E*,4*E*)-Dec	(*S*)-2MeBu	H	OH	*F. picrosperma*	[8]
134	(2*E*,4*E*,6*E*)-Dec	Ac	H	OH	*Ex. acerifolia*	[127]
135	(2*E*,4*E*,6*E*)-Dodec	(*S*)-2MeBu	H	OH	*F. picrosperma*	[8]
136	(*S*)-2MeBu	(*S*)-2MeBu	(9Z)-Octadec	OH	*F. picrosperma*	[36]
137	(*S*)-2MeBu	(*S*)-2MeBu	(9Z,12Z)-Octadec	OH	*F. picrosperma*	[36]
138	Tigl	(*S*)-2MeBu	(9Z,12Z)-Octadec	OH	*F. picrosperma*	[36]
139	Hex	(*S*)-2MeBu	(9Z)-Octadec	OH	*F. picrosperma*	[36]
140	Hex	(*S*)-2MeBu	(9Z,12Z)-Octadec	OH	*F. picrosperma*	[36]
141	(2*E*,4*E*)-Dec	(*S*)-2MeBu	(9Z)-Octadec	OH	*F. picrosperma*	[36]
142	(2*E*,4*E*)-Dec	(*S*)-2MeBu	(9Z,12Z)-Octadec	OH	*F. picrosperma*	[36]
143	(2*E*,4*E*,6*E*)-Dodec	(*S*)-2MeBu	(9Z)-Octadec	OH	*F. picrosperma*	[36]
144	(2*E*,4*E*,6*E*)-Dodec	(*S*)-2MeBu	(9Z,12Z)-Octadec	OH	*F. picrosperma*	[36]

Ex. = Excoecaria; F. = Fontainea; Sa. = Sapium
*Most probably a β-epoxide and an isolation artefact

di- and triesters identified so far belong to the A- and C-type, while the ester group at C-20 is an unsaturated fatty acid (oleic or linoleic) (Table 4; **3**, **125–144**). Interestingly, the lipid matrix hosting these compounds is mostly made up by saturated fatty acids, suggesting a precise, yet elusive, acyl decoration strategy, fully independent from the hosting fatty acids pool.

4-Deoxyphorbol Esters

4-Deoxyphorbol Esters Sensu Stricto

The deletion of the 4-hydroxy group is detrimental for stability toward epimerization at C-4, and it is therefore not surprising that many 4-deoxyphorbol esters (Table 5, **145–183**) have been reported to co-occur with their isomers from the 4α-series, which are therefore isolation artefacts, belonging to the programmed and not the native chemical diversity. Type-A and type-B polyunsaturated esters co-occur in *E. tirucalli* (Plate 5), the most prolific source of this type of phorbol ester. Of relevance is the occurrence of type-C *ambo-N,N*-dimethylvaline esters, which are esters having enantiomeric *N,N*-dimethylvaline residues at the 12-hydroxy group.

Mono- and Dihydroxylated 4-Deoxyphorbol Esters

This rare group of phorbol esters **184–191** is documented within plants from the genera *Sapium* and *Croton* (Table 6). The occurrence of a type-C ester hydroxylated at C-17 from *Euphorbia obtusifolia* (compound **190**) is unclear in terms of plant origin [131], since this plant name can refer to three distinct species, of which two are present in the location of collection (Canary islands, *Euphorbia lamarckii* Sweet and *Euphorbia regis-jubae* J.Gay). On the other hand, the oxygenation of the β-oriented C-17 methyl was confirmed by the detection of a diagnostic ^{1}H NMR NOE-effect between the 11β-proton, and appears therefore to be correct [131]. Compounds **110** [113] and **190** [131] therefore seem to be genuine C-17 esters.

12-Deoxyphorbol Esters

Esters of 12-deoxyphorbol are the largest group of naturally occurring tigliane derivatives, and are especially frequent in plants of the genus *Euphorbia*, although they are not generally accumulated to a significant extent when compared to other classes of diterpenoids like the jatrophanes.

12-Deoxyphorbol Esters Sensu Stricto

These compounds occur frequently in euphorbiaceous plants, but have been isolated from a single thymelaeceous genus (*Pimelea*, Table 7). Within the various euphorbiaceous genera, they are remarkably absent in plants from the genus *Croton*, of which the members are prolific producers of phorboids.

Prostratin (13-acetyl-12-deoxyphorbol) (**192**) is one of the most widespread tigliane esters among **192–257**, having been isolated from eleven different euphorbiaceous species belonging to five distinct genera, with its 20-glucoside (**224**) and galloylated glucosides (**228–230**) having additionally been reported from four species.

Table 5 Naturally occurring 4-deoxyphorbol esters

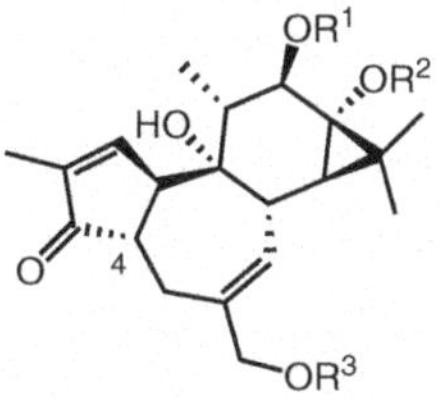

Compound	R1	R2	R3	Source	References
145	H	H	H	*Sa. hippomane*	[128]
146	H	Ac	H	*E. tirucalli*	[129]
147	Ac	Δ^{2}-Oct	H	*E. tirucalli*	[130]
148	Ac	(2*Z*,4*E*)-Oct	H	*E. tirucalli*	[130]
149	Ac	$\Delta^{2,4}$-Dec	H	*E. tirucalli*	[130]
150	Ac	*(2Z,4E)*-$\Delta^{2,4,6}$-Dec	H	*E. tirucalli*	[130]
151	Ac	$\Delta^{2,4,6}$-Dec	H	*E. tirucalli*	[130
152	Ac	$\Delta^{2,4,6}$-Dodec	H	*E. tirucalli*	[130]
153	Ac	$\Delta^{2,4,6,8}$-Dodec	H	*E. tirucalli*	[130]
154	Ac	$\Delta^{2,4,6,8}$-Tetradec	H	*E. tirucalli*	[130]
155	Ac	$\Delta^{2,4,6,8,10}$-Tetradec	H	*E. tirucalli*	[130]
156	*i*Bu	*i*Bu	H	*E. obtusifolia*	[131]
157	Tigl	Ac	H	*C. tiglium*	[132]
158	Tigl	*i*Bu	H	*Syn. grantii*	[133]
159	Tigl	2MeBu	H	*Sa. hippomane*	[128]
160	Tigl	PhAc	H	*Syn. grantii* *E. umbellata*	[134] [135]
161	Tigl	Hexadec	H	*C. tiglium*	[132]
162	Sen	Ac	H	*C. ciliatoglandulifer*	[94]
163	D-*N*,*N*-dimethylvaline	Ac	H	*C. ciliatoglandulifer*	[94]
164	L-*N*,*N*-dimethylvaline	Ac	H	*C. ciliatoglandulifer*	[94]
165	Me-2Abz-OH	Ac	H	*Sa. indicum* *Sa. sebiferum*	[76, 109, 115, 137] [136]
166	(2*Z*,4*E*)-Oct	Ac	H	*E. tirucalli*	[129]
167	$\Delta^{2,4}$-Oct	Ac	H	*E. tirucalli*	[130]
168	(2*Z*,4*E*)-$\Delta^{2,4,6}$-Dec	Ac	H H	*E. tirucalli* *Sa. indicum*	[130] [115, 138]
169	$\Delta^{2,4,6}$-Dec	Ac	H	*E. tirucalli*	[130]
170	$\Delta^{2,4,6,8}$-Dodec	Ac	H	*E. tirucalli*	[130]
171	$\Delta^{2,4,6,8}$-Tetradec	Ac	H	*E. tirucalli*	[130]
172	$\Delta^{2,4}$-Dec	*i*Bu	H	*E. prolifera*	[139]

(continued)

Table 5 (continued)

Compound	R^1	R^2	R^3	Source	References
173	$\Delta^{2,4,6}$-Dec	*i*Bu	H	*E. prolifera*	[139]
174	(2*E*,4*E*,6*E*,8*E*)-Tetradec	*i*Bu	H	*E. tirucalli*	[129]
175	H	Ac	(9*Z*)-Octadec	*C. tiglium*	[132]
176	H	Ac	(9*Z*,12*Z*)-Octadec	*C. tiglium*	[132]
177	Ac	Ac	Ac	*Sa. hippomane*	[128]
178	*i*Bu	*i*Bu	Ac	*E. obtusifolia*	[131]
179	*i*Bu	Ac	Bz	*E. nubica*	[140]
180	Bz	*i*Bu	Ac	*E. nubica*	[140]
181	(2*Z*,4*E*)-Oct	Ac	Ac	*E. broteri* *E. biglandulosa*	[120] [141]
182	(2*Z*,4*E*)-Oct	Pr	Ac	*E. biglandulosa*	[141]
183	(2*Z*,4*E*)-Oct	*i*Bu	Ac	*E. biglandulosa*	[141]

C. = Croton; E. = Euphorbia; Sa. = Sapium, Syn.= Synadeniun

Plate 5 *Euphorbia tirucalli* L. Photograph courtesy Nevit Dilmen; Creative Commons 3.0

Table 6 Naturally occurring mono- and dihydroxylated 4-deoxyphorbol esters

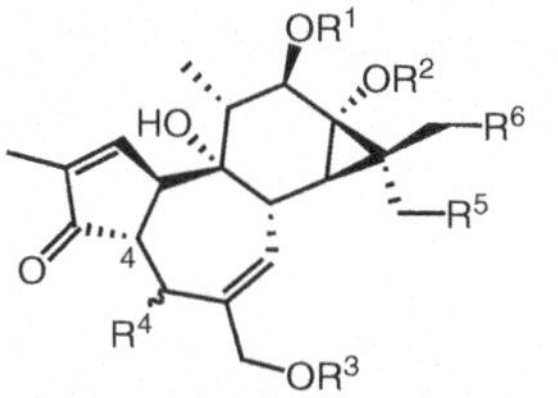

Compound	R^1	R^2	R^3	R^4	R^5	R^6	Source	References
184	H	H	H	H	OAc	H	*S. insigne*	[109]
185	Me-2Abz-OH	Ac	H	*β*-OH	H	H	*S. indicum*	[109, 115, 137]
186	(2*Z*,4*E*)-$\Delta^{2,4,6}$-Dec	Ac	H	*β*-OH	H	H	*S. indicum*	[115, 138]
187	(2*E*,4*E*)-Dec	Ac	H	H	OH	H	*S. insigne*	[109]
188	Hexadec	Ac	H	H	OH	H	*C. flavens*	[124]
190	*i*Bu	*i*Bu	H	H	H	OAc	*E. obtusifolia*	[131]
191	Hexadec	Ac	Dec	H	OH	H	*C. flavens*	[124]

C.= Croton; E. = Euphorbia; S. = Sapium

Hydroxylated 12-Deoxyphorbol Esters

The additional hydroxylation sites of 12-deoxyphorbol derivatives are the same as phorbol (**1**), namely, the C-5 methylene and the C-16 methyl (Table 8). The *Jatropha* factors **17–21** [68] are of special interest among **258–284**, since the 12- and 16-hydroxy groups are esterified by single bicarboxylic acids, for which their cyclobutane and bicyclo[3.1.0]hexane structures could derive from the dimerization of two distinct polyunsaturated acids located at C-13 and C-16, as discussed in Sect. 3.3. Just like its corresponding phorbol derivative, 12-deoxy-16-hydroxyphorbol is also degraded during hydrolysis of its esters, and the parent polyol has never been characterized. Of biogenetic relevance is the co-occurrence of the daphnane orthoester resiniferatoxin (**10**) and tigliane 16-esters, due to a possible biogenetic relationship between diterpenoids from the tigliane and the daphnane classes (see Sect. 4).

Epoxidized 12-Deoxyphorbol Esters

In close similarity to observations with phorbol esters, the enzymatic epoxidation of the ring B double bond apparently requires the presence of a 5*β*-hydroxy function. Although few in number, esters of this type (**285–292**) have been reported from six distinct euphorbiaceous species and one thymelaeaceous plant (Table 9).

20-Deoxyphorbol Esters

This small group of esters (**293–297**) (Table 10) is remarkable in the presence of 5-hydroxylated compounds where an *α*- and not the usual *β*-configuration has been

Table 7 Naturally occurring 12-deoxy phorbol esters

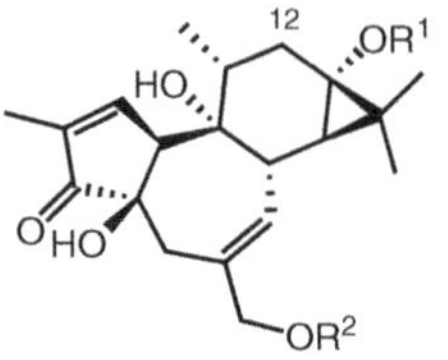

Compound	R[1]	R[2]	Source	References
192	Ac	H	*E. stracheyi* *H. nutans* *E. fischeriana* *P. simplex* *P. prostrata* *E. triangularis* *E. cornigera* *E. papillosa* *Sa. glandulosum* *St. sylvatica* *D. racemosa* *E. kansuensis*	[76, 78] [142] [80, 116, 143, 144] [145] [145] [146] [147] [148] [148] [149] [150] [83]
193	*i*Bu	H	*Sy. grantii* *E. ledienii* *E. resinifera* *E. triangularis* *E. poissonii*	[118] [146] [151] [146, 152] [153]
194	2MeBu	H	*Stil. lineata* *E. coerulescens* *E. ledienii* *E. papillosa* *Sa. glandulosum*	[86] [146] [146] [148] [148]
195	(*S*)-2MeBu	H	*E. triangularis* *E. poissonii* *E. unispina* *E. coerulescens*	[152] [153] [153] [154]
196	Ang	H	*E. resinifera* *E. unispina* *E. poissonii*	[152] [153] [153, 155]
197	Tigl	H	*E. ledienii* *E. triangularis* *E. fortissima* *E. helioscopia*	[146] [152] [154] [146, 154]
198	Hept	H	*E. coerulescens*	[154]
199	Bz	H	*Sa. sebiferum*	[136]
200	Oct	H	*Ne. melleri* *A. malaccensis*	[156] [95]
201	PhAc	H	*E. papillosa* *Sa. glandulosum* *E. resinifera* *E. poissonii*	[148] [148] [151] [157]
202	4-(OMe)PhAc	H	*E. resinifera*	[151]
203	Dec	H	*E. fischeriana* *Al. malaccensis*	[145] [95]

(continued)

Table 7 (continued)

Compound	R^1	R^2	Source	References
204	*(3E,5E)*-Dec	H	*Ex. agallocha*	[158]
205	(9*S*,10*S*-Me)-Undec	H	*E. poissonii*	[155]
206	Dodec	H	*E. coerulescens* *Stil. loranthacea* *Re. trisperma*	[154] [159] [160]
207	Dodec:1	H	*E. fortissima*	[154]
208	Tetradec	H	*E. stracheyi* *Sa. sebiferum* *Stil. loranthacea*	[76, 78] [109] [159]
209	12-(Me)Tetradec	H	*E. fischeriana*	[144]
210	(5*Z*)-Tetradec	H	*Stil. loranthacea*	[159]
211	$\Delta^{3,5}$-Tetradec	H	*Ex. agallocha*	[161]
212	Tetradec:1	H	*Sa. sebiferum*	[109]
213	Tetradec:2	H	*Sa. sebiferum*	[109]
214	Tetradec:3	H	*Sa. sebiferum*	[109]
215	(2*E*,4*E*)-6-OH-Tetradec	H	*E. acertifolia*	[85]
216	Pentadec	H	*Ne. macrocalyx*	[162]
217	Hexadec	H	*Sa. sebiferum* *E. fischeriana* *Re. trisperma* *B. montanum* *E. kansuensis*	[109] [116, 144, 163, 164] [160] [165] [83]
218	(7*Z*)-Hexadec	H	*E. fischeriana*	[164]
219	$\Delta^{3,5,7}$-Hexadec	H	*Ex. agallocha*	[161]
220	(2*E*,4*E*,6*E*)-8-Oxo-hexadec	H	*Stil. lineata*	[86]
221	(6*Z*)-Octadec	H	*E. fischeriana*	[164]
222	(9*Z*,12*Z*)-Octadec	H	*E. fischeriana*	[164]
223	Ac	Ac	*E. fischeriana* *E. kansuensis*	[116, 144] [83]
224	Ac	*β*-D-Glu	*E. fischeriana* *E. ebracteolata* *E. stracheyi* *E. kansuensis*	[80, 82 [81] [76] [83]
225	Ac	Bz	*E. cornigera*	[147]
226	Ac	4OMeBz	*E. cornigera*	[147]
227	Ac	Dec	*E. cornigera*	[147]
228	Ac	3-Gall-*β*-D-Glu	*E. fischeriana*	[80]
229	Ac	4-Gall-*β*-D-Glu	*E. fischeriana*	[80]
230	Ac	6-Gall-*β*-D-Glu	*E. fischeriana*	[116]
231	Ac	(9*Z*)-Octadec	*Al. malaccensis*	[95]
232	Ac	(9*Z*,12*Z*)-Octadec	*E. kansuensis*	[83]
233	Ac	(9*Z*,12*Z*,15*Z*)-Octadec	*E. kansuensis*	[83]
234	*i*Bu	Ac	*E. ledienii* *E. triangularis* *E. papillosa* *Sa. glandulosum* *E. resinifera* *E. poissonii* *E. fortissima*	[146] [146, 152] [148] [148] [151] [153] [154]

(continued)

Table 7 (continued)

Compound	R^1	R^2	Source	References
235	Bu	Dec	*E. cornigera*	[147]
236	2MeBu	Ac	*E. ledienii* *E. triangularis* *E. coerulescens*	[146] [146] [146]
237	(*S*)-2MeBu	Ac	*E. triangularis* *E. poissonii* *E. unispina* *E. coerulescens* *E. fortissima*	[152] [153] [153] [154] [154]
238	Ang	Ac	*E. resinifera* *E. unispina* *E. poissonii*	[151] [153] [153, 155]
239	Tigl	Ac	*E. ledienii* *E. helioscopia* *E. triangularis* *E. coerulescens* *E. fortissima*	[146] [146, 154] [152] [154] [154]
240	Hex	Dec	*E. cornigera*	[147]
241	Oct	Dec	*E. cornigera*	[147]
242	Oct:1	Ac	*E. polyacantha*	[154]
243	PhAc	Ac	*E. papillosa* *Sa. glandulosum* *E. resinifera* *E. poissonii*	[147] [148] [151] [155, 157]
244	4OHPhAc	Ac	*E. poissonii*	[157]
245	4OMePhAc	Ac	*E. resinifera*	[151]
246	Dec	Ang	*E. cornigera*	[147]
247	Dec	Tigl	*E. cornigera*	[147]
248	Dec	Dec	*E. cornigera*	[147]
249	Dec:2	Ac	*E. polyacantha*	[154]
250	4OAcPhAc	Ac	*E. poissonii*	[157]
251	Dodec	Dec	*E. cornigera*	[147]
252	Dodec	Hexadec	*Ne. macrocalyx*	[162]
253	Dodec	Ac	*E. coerulescens*	[154]
254	Dodec:1	Ac	*E. fortissima*	[154]
255	Dodec:2	Ac	*E. helioscopia*	[154]
256	Tetradec	Ac	*E. stracheyi* *Al. moluccana*	[76, 78] [166]
257	(9*Z*)-Octadec	Ac	*E. fischeriana*	[145]

Al. = Aleurites; B. = Baliospermum; E. = Euphorbia; E.x = Excoecaria; Ne. = Neobutonia; O. = Ostodes; P. = Pimelea; Re. = Reuteralis; Sa. = Sapium; Stil. = Stillingia; Sy. = Synademium

assigned to this stereocenter. This structural assignment should be confirmed by modern NMR studies, but it is interesting that the assignment of an α-configuration to the 5-hydroxy group has always been made in compounds where C-20 is a methyl and not an oxymethylene, suggesting subtle effects in the specificity of the enzymatic

Table 8 Naturally occurring 5(16)-hydroxy-12-deoxyphorbol esters

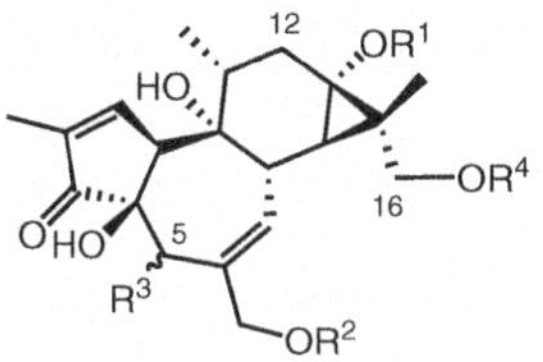

Compound	R^1	R^2	R^3	R^4	Source	References
17	—R_a—	H	H	—R_a—	*J. curcas*	[68]
18	—R_b—	H	H	—R_b—	*J. curcas*	[68]
19	—R_c—	H	H	—R_c—	*J. curcas*	[68]
20*	—R_d—	H	H	—R_d—	*J. curcas*	[68]
21	—R_e—	H	H	—R_e—	*J. curcas*	[68]
258	H	H	H	Ac	*E. kansuensis*	[83]
259	Ac	H	*β*-OH	H	*E. acertifolia* *D. racemosa*	[85] [150]
260	Tetradec	H	*β*-OH	H	*B. montanum*	[165]
261	(2*E*,4*E*)-6-OH-Tetradec	H	*β*-OH	H	*E. acertifolia*	[85]
262	Hexadec	H	*β*-OH	H	*Sti. sylvatica*	[149]
263	Hexadec	H	H	H	*B. montanum*	[165]
264	Ac	H	H	*β*-D-Glu	*E. ebracteolata*	[81]
265	Ang	H	H	*i*Bu	*E. ledienii* *E. cooperi* *E. triangularis*	[146] [146, 162] [146]
266	Ang	H	H	2MeBu	*E. coerulescens*	[146]
267	Tigl	H	H	*i*Bu	*E. cooperi*	[167]
268	PhAc	H	H	*i*Bu	*E. resinifera*	[168]
269	PhAc	H	H	(*S*)-2MeBu	*E. poissonii*	[157]
270	PhAc	H	H	Tigl	*E. resinifera*	[168]
271	PhAc	H	H	Bz	*E. resinifera*	[151]
272	Dec	H	H	2MeBu	*C. megalobotrys*	[169]
273	Dodec	H	H	2MeBu	*C. megalobotrys*	[169]
274	(3*E*,5*E*)-Tetradec	H	H	Ac	*Sa. insigne*	[109]
275	(*R*,2*E*,4*E*)-6-OH-Tetradec	H	*β*-OH	Ac	*Sa. insigne*	[109]
276	(*S*,2*E*,4*E*)-6-OH-Tetradec	H	*β*-OH	Ac	*Sa. insigne*	[109]

(continued)

Table 8 (continued)

Compound	R^1	R^2	R^3	R^4	Source	References
277	(2*E*,4*E*)-6-O-Tetradec	H	*β*-OH	Ac	*Sa. insigne*	[109]
279	*i*Bu	Ac	H	Ang	*E. cooperi*	[170]
280	Tigl	Ac	H	*i*Bu	*E. cooperi*	[167]
281	Ang	Ac	H	*i*Bu	*E. ledienii* *E. cooperi* *E. triangularis*	[146] 138 [146]
282	Ang	Ac	H	2MeBu	*E. ledienii* *E. coerulescens*	[146] [146]
283	PhAc	Ac	H	(*S*)-2MeBu	*E. poissonii*	[157]
284	PhAc	Ac	H	Bz	*E. resinifera*	[151]

B. = Baliospermum; C. = Croton; D. = Daphne; E. = Euphorbia; J. = Jatropha; Sa. = Sapium; Sti. = Stillingia
**Mixture of stereoisomers*

Table 9 Naturally occurring 6*α*,7*α*-epoxy-12-deoxy phorbol esters

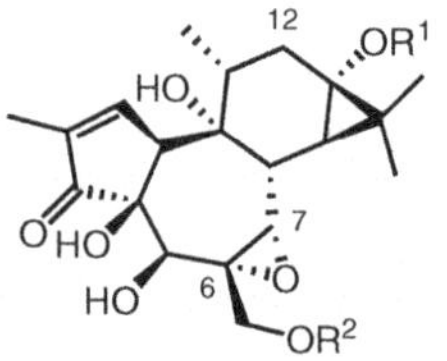

Compound	R^1	R^2	Source	References
285	Ac	H	*E. acertifolia*	[85]
286	Dec	H	*P. elongata* *W. chamaedaphne*	[103] [171]
287	Dodec	H	*B. montanum*	[165]
288	Tetradec	H	*P. simplex* *P. prostrata* *Ne. macrocalyx*	[145] [145] [162]
289	$\Delta^{2,4}$-Tetradec	H	*P. simplex*	[172]
290	(2*E*,4*E*)-Tetradec	H	*P. elongata*	[103]
291	(2*E*,4*E*,6*E*)-Hexadec	H	*Hip. mancinella*	[173]
292	Hexadec	Dodec	*Ne. macrocalyx*	[162]

B. = Baliospermum; E. = Euphorbia; Hip. = Hippomane; Ne. = Neobutonia; P. = Pimelea; W. = Wikstroemia

Table 10 Naturally occurring 20-deoxy phorbol esters

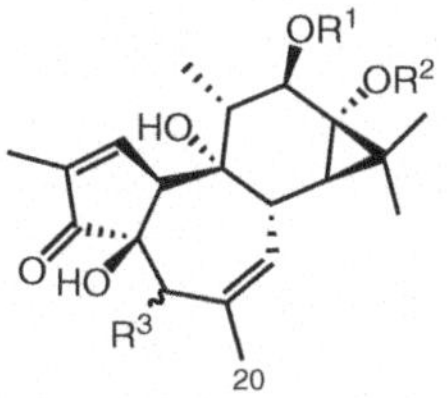

Compound	R^1	R^2	R^3	Source	References
293	Tigl	*i*Bu	α-OH	*Sy. grantii*	[139]
294	Tigl	*i*Bu	β-OH	*E. umbellata*	[135]
295	Tigl	PhAc	α-OH	*Sy. grantii*	[139]
296	Tigl	PhAc	β -OH	*E. umbellata*	[135]
297	Bz	*i*Bu	α-OH	*E. neorubella*	[174]

E. = *Euphorbia; Sy.* = *Synadenium*

Table 11 Naturally occurring 4,12-dideoxy phorbol esters

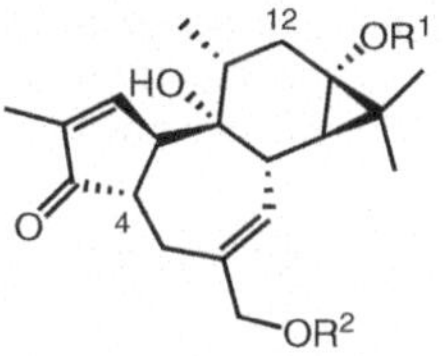

Compound	R^1	R^2	Source	References
298	2,3MeBu	H	*E. pithyusa* spp. *cupanii*	[175, 176]
299	$\Delta^{3,5}$-Tetradec	H	*Ex. bicolor*	[177]
300	*i*Bu	Bz	*E. pannonica*	[178]
301	3MeBu	Bz	*E. pannonica*	[178]
302	2,3MeBu	Ac	*E. pithyusa* spp. *cupanii*	[175]

E. = *Euphorbia; Ex.* = *Excoecaria*

epoxidation, or, alternatively, the existence of distinct enzymes with distinct substrate specificity for the reaction.

4,12-Dideoxyphorbol Esters

Tigliane diesters (**298–302**) (Table 11) from this polyol are significantly less polar than other tigliane classes. Due to their lipophilicity and difficulty of isolation from fatty matrices, their occurrence might have been underestimated.

Table 12 Naturally occurring 4,20-dideoxy phorbol esters

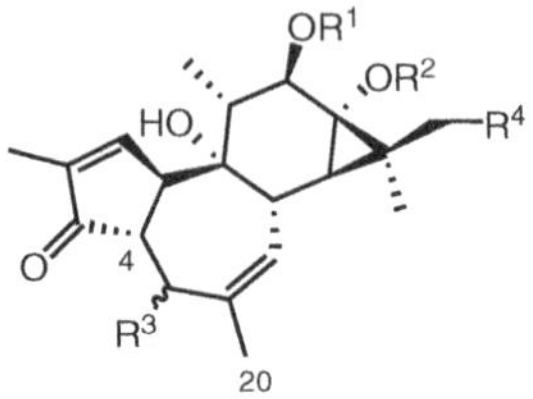

Compound	R^1	R^2	R^3	R^4	Source	References
303	*i*Bu	*i*Bu	H	H	*E. obtusifolia*	[131]
304	*i*Bu	*i*Bu	OH	H	*E. prolifera*	[139]
305	*i*Bu	*i*Bu	*β*-OH	H	*E. tuckeyana* *E. neorubella*	[179] [174]
306	Tigl	*i*Bu	*α*-OH	H	*Sy. grantii*	[134]
307	Tigl	PhAc	*α*-OH	H	*Sy. grantii*	[134]
308	Bz	*i*Bu	H	H	*E. prolifera* *E. sanctae-catharinae*	[139] [180]
309	Bz	*i*Bu	OH	H	*E. prolifera*	[139]
310	Bz	*i*Bu	*β*-OH	H	*E. tuckeyana* *E. neorubella* *E. dracunculoides* *E. prolifera*	[179] [174] [181] [182]
311	Me-2Abz-OH	Ac	H	H	*Sa. indicum*	[109, 137]
312	Me-2Abz-OH	Ac	*β*-OH	H	*Sa. indicum* *Sa. sebiferum*	[109, 115, 137] [136]
313	(2*Z*,4*E*)-$\Delta^{2,4,6}$-Dec	Ac	*β*-OH	H	*Sa. indicum*	[115, 138]
314	Tetradec	Ac	*β*-OH	H	*Sa indicum*	[115]
315	Bz	*i*Bu	OAc	H	*E. nubica*	[140]
316	*i*Bu	*i*Bu	H	OAc	*E. obtusifolia*	[131]

E. = Euphorbia; Sa = Sapium; Sy = Synadenium

4,20-Dideoxyphorbol Esters

Some unusual structural features have been reported in this class of esters (**303–316**), namely, functionalization of C-17 and not C-16, and 5*α*-hydroxylation rather than 5*β*-hydroxylation (Table 12). This feature seems to be associated with the lack of oxygenation of C-20, as also observed in 20-deoxyphorbol derivatives.

12,20-Dideoxyphorbol Esters

Euphorbia resinifera Berg. (Plate 6) is so far the only source of this type of compound, representing one of the rarest types of phorbol esters typified by **317** and **318** (Table 13).

Plate 6 *Euphorbia resinifera* Berg. Photograph courtesy Valery & Agnes; Creative Commons 3.0

Table 13 Naturally occurring 12,20-dideoxy phorbol esters

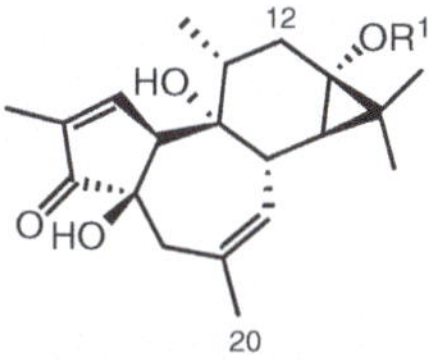

Compound	R^1	Source	References
317	*i*Bu	*E. resinifera*	[151]
318	Ang	*E. resinifera*	[151]

E. = *Euphorbia*

Trideoxy Phorbol Esters

These compounds are the least functionalized class of tigliane esters, and only one compound, **319**, of this type has so far been reported, from *Euphorbia pithyusa* L. subsp. *cupanii* (Guss. ex Bertol.) Radcl.-Sm. (Plate 7), of Sardo-Corsican and Sicilian endemism (Table 14). The retaining of combined hydroxylation at C-9 and C-13 is

Plate 7 *Euphorbia pithyusa* L. subsp. *cupanii* (Guss. ex Bertol.) Radel.-Sm. Photograph courtesy Ghislain118; Creative Commons 2.5

Table 14 Naturally occurring 4,12,20-trideoxy phorbol esters

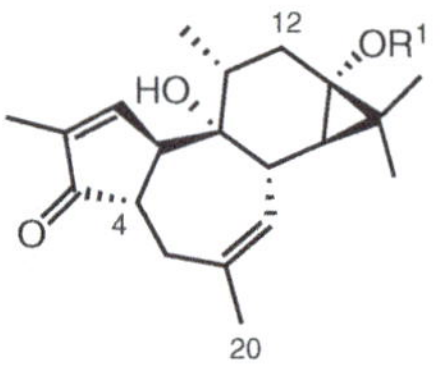

Compound	R_1	Source	References
319	2,3MeBu	*E. pithyusa* ssp. *cupanii*	[175]

E. = *Euphorbia*

in accordance with the generation of functionalization at these carbons in a single biosynthetic step, as postulated by Jakupovic (see Sect. 3.4).

Table 15 Naturally occurring 3-dihydrophorbol esters

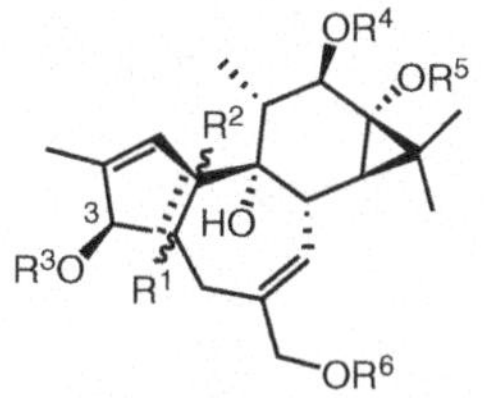

Compound	R^1	R^2	R^3	R^4	R^5	R^6	Source	References
320	*β*-OH	*α*-H	H	H	H	H	*E. fischeriana*	[183]
321	–		H	H	H	H	*E. fischeriana*	[183]
322	*β*-OAc	*α*-H	Ac	Ac	Ac	PhAc	*Sy. grantii*	[184, 185]

E. = *Euphorbia; Sy* = *Synadenium*

Functional Group-Modified Phorbol Esters

3-Dihydrophorbol Esters

Three compounds (**320–322**) (Table 15) of this type have been reported, with two of these being remarkable for, respectively, the presence of a 4,10-double bond, a structural apex among tigliane diterpenoids, and for a tetraester structure, with acetylation also at the tertiary 4-hydroxy group. In all cases, the secondary 3-hydroxy group is *β*-oriented, a feature reminiscent of ingenol esters.

C-11 Functionalized Phorbol Esters

This very rare class of esters includes compound **323** that is hydroxylated at C-11 and was isolated from *Euphorbia fischeriana* Steud. [80] (Table 16), and also three compounds (**22–24**, Table 17) with a $\Delta^{11,18}$ unsaturation, as purified from the African plant *Pycnocoma cornuta* Muell. [69]. These three compounds are interesting also for the presence of a complex bicyclo[3.1.0]-based dicarboxylic acid related to that present in the *Jatropha* factors, and which presumably has a similar biogenesis but with hydroxylated cinnamate acids replacing the C-10 component (see Sect. 3.2).

Ring B Oxidatively Modified Phorbol Esters

The presence of a hydroperoxy group at C-8 in compound **324** (Table 18) is an unusual feature, since the ene-reaction of a putative $\Delta^{7,8}$-bond would generate a 7-hydroperoxide with a strained $\Delta^{8,14}$-bond (see Sect. 4.2.1). On the other hand, an auto-oxidation reaction would involve a rearrangement-prone cyclopropylmethyl radical and opening of the D-ring. The epoxide **327** might be an artefact resulting from the further oxidation of an initial 7-hydroperoxy-$\Delta^{5,6}$ auto-oxidation product.

Table 16 Naturally occurring 9α-9,12-dideoxy phorbol ester

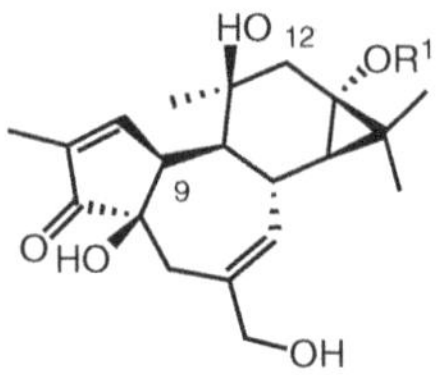

Compound	R^1	Source	References
323	Ac	*E. fischeriana*	[80]

E. = *Euphorbia*

Table 17 Naturally occurring 11,18-dehydro phorbol esters

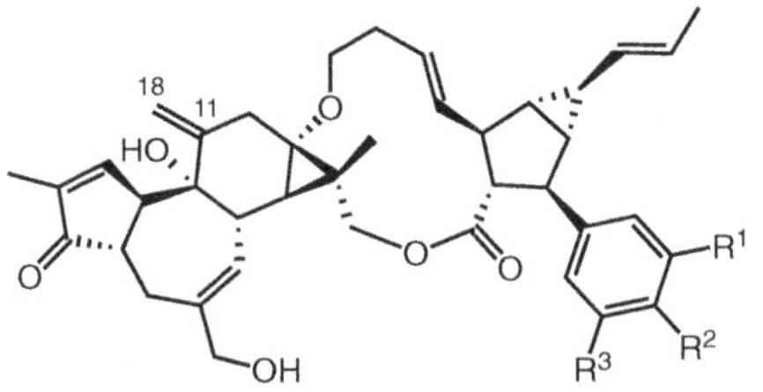

Compound	R^1	R^2	R^3	Source	References
22	H	OMe	H	*Pyc. cornuta*	[69]
23	OMe	OMe	H	*Pyc. cornuta*	[69]
24	–OCH_2O–		OMe	*Pyc. cornuta*	[69]

Pyc. = *Pycnocoma*

Table 18 Naturally occurring ring B oxidatively modified phorbol esters

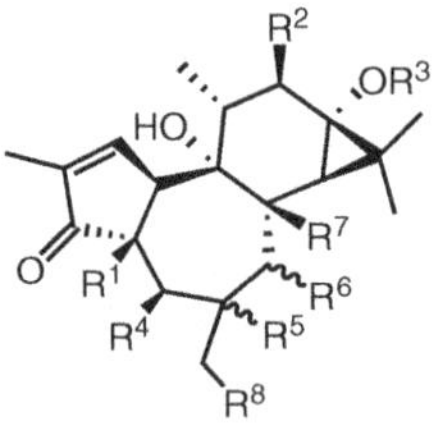

Compound	R^1	R^2	R^3	R^4	R^5	R^6	R^7	R^8	Source	References
324	H	OBz	*i*Bu	OH	Δ		OOH	H	*E. dracunculoides*	[186]
325	OH	H	Dec	(β)-O-		α-OH	H	OH	*W. chamaedaphne*	[171]
326	OH	H	Tetradec	H	β-OH	β-OH	H	OH	*R. trisperma*	[160]
327	OH	OHexadec	Ac	H	(β)-O-		H	OH	*Sa. sebiferum*	[126]

E. = *Euphorbia; R.* = *Reuteralis ; Sa.* = *Sapium; W.* = *Wikstroemia*

Table 19 Naturally occurring Δ^7-phorbol esters

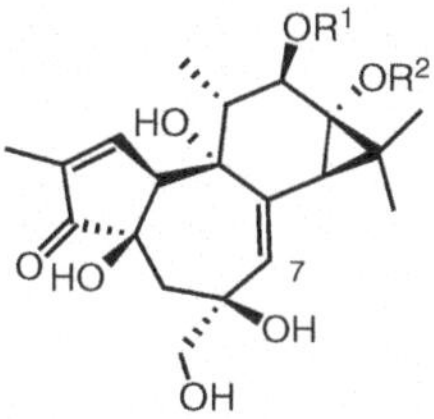

Compound	R^1	R^2	Source	References
328	Bz	Oct	*W. lamatsoensis* *W. scytophylla*	[99] [123]

W. = *Wikstroemia*

Δ^7-Phorbol Esters

The 12,13-diester **328** (Table 19), isolated from plants of the thymelaeaceous *Wikstroemia* genus, is the only natural phorbol derivative with a Δ^7-double bond reported so far [99, 123]. The Δ^7-double bond could possibly derive from the dehydration of a 7-functionalized precursor, a reaction documented in the chemical manipulation of phorbol esters (see Sect. 4.1). A possible biogenetic scheme could involve an ene-reaction with hydrogen abstraction by oxygen from the allylic C-20 methyl, rather than the allylic C-5 methylene, followed by dihydroxylation of the $\Delta^{6,20}$-double bond and elimination of the C-7 oxygen function.

Possible Isolation Artefacts of Phorbol Esters

Phorbol esters occur as complex mixtures of related compounds, and extensive chromatographic activity is required for their purification. This, coupled with the propensity of the natural products to undergo isomerization reactions and to undergo autooxidation, explains the abundance of artefacts plaguing the tigliane literature.

4α-4-Deoxyphorbol Esters

Due to the very easy isomerization of 4-deoxyphorbol to 4*α*-4-deoxyphorbol (see Sect. 4.1), these compounds (**329–371**; Table 20) could have originated from their corresponding native deoxyphorbol esters during isolation. The co-occurrence of the two types of esters strongly supports this point of view.

Hydroxylated 4α-4-Deoxyphorbol Esters

The presence of hydroxylation at the geminal methyls at C-14 in **373–377** does not affect the retroaldol equilibration of the A/B-ring system, but it would be interesting to confirm the *β*-configuration of the 5-hydroxy group in ester **372** (Table 21), since

Table 20 4α-4-Deoxy phorbol esters (possibly isolation artefacts)

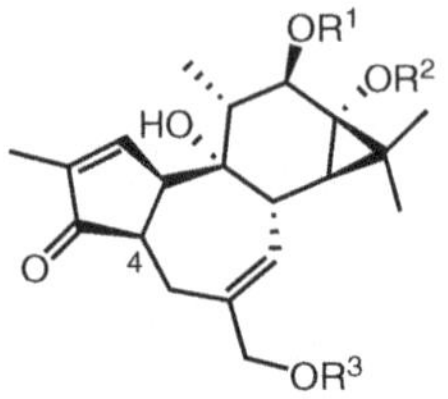

Compound	R^1	R^2	R^3	Source	References
329	H	Ac	H	*E. tirucalli*	[129]
330	H	*i*Bu	H	*C. damayeshu*	[187]
331	H	(*S*)-2MeBu	H	*C. damayeshu*	[187]
332	H	(2*S*,3*S*)-3-OH-2MeBu	H	*C. damayeshu*	[187]
333	Ac	(*S*)-2MeBu	H	*Stil. lineata*	[86]
334	Ac	Δ^{2}-Oct	H	*E. tirucalli*	[130]
335	Ac	$\Delta^{2,4}$-Oct	H	*E. tirucalli*	[130]
336	Ac	$\Delta^{2,4}$-Dec	H	*E. tirucalli*	[130]
337	Ac	$\Delta^{2,4,6}$-Dec	H	*E. tirucalli*	[130]
338	Ac	$\Delta^{2,4,6}$-Dodec	H	*E. tirucalli*	[130]
339	Ac	$\Delta^{2,4,6,8}$-Dodec	H	*E. tirucalli*	[130]
340	Ac	$\Delta^{2,4,6,8}$-Tetradec	H	*E. tirucalli*	[130]
341	Ac	$\Delta^{2,4,6,8,10}$-Tetradec	H	*E. tirucalli*	[130]
342	Bu	(*S*)-2MeBu	H	*C. damayeshu*	[187]
343	*i*Bu	*i*Bu	H	*E. obtusifolia*	[131]
344	2MeBu	*i*Bu	H	*E. sogdiana*	[188]
345	(*S*)-2MeBu	*i*Bu	H	*C. damayeshu*	[187]
346	(*S*)-2MeBu	(*S*)-2MeBu	H	*C. damayeshu*	[187]
347	2,3MeBu	*i*Bu	H	*E. aelleni* *E. sogdiana*	[189] [190]
348	Ang	(*S*)-2MeBu	H	*C. damayeshu*	[187]
349	Tigl	Ac	H	*C. tiglium* *C. damayeshu*	[89, 92, 191] [187]
350	Tigl	*i*Bu	H	*C. tiglium* *C. damayeshu*	[89] [187]
351	Tigl	2MeBu	H	*C. tiglium* *Sa. hippomane*	[89] [128]
352	Tigl	(*S*)-2MeBu	H	*C. damayeshu*	[187]
353	Tigl	(2*S*,3*S*)-3-OH-2MeBu	H	*C. damayeshu*	[187]
354	Tigl	PhAc	H	*C. tiglium*	[89]
355	Tigl	Dec	H	*C. tiglium*	[89]
356	Hex	Ac	H	*Stil. insigne*	[109]
357	Bz	Ac	H	*C. tiglium*	[92]
358	Bz	(*S*)-2MeBu	H	*C. damayeshu*	[187]
359	Me-2Abz-OH	Ac	H	*Sa. indicum*	[109, 115, 137]

(continued)

Table 20 (continued)

Compound	R¹	R²	R³	Source	References
360	(2*Z*,4*E*)-Oct	Ac	H	*E. tirucalli* *Pe. tithymaloides*	[129] [192]
361	(2*Z*,4*E*,6*E*)-Non	Bu	H	*Stil. lineata*	[86]
362	(2*Z*,4*E*)-$\Delta^{2,4,6}$-Dec	Ac	H	*E. tirucalli*	[130]
363	$\Delta^{2,4,6}$-Dec	Ac	H	*E. tirucalli*	[130]
364	$\Delta^{2,4,6,8}$-Dodec	Ac	H	*E. tirucalli*	[130]
365	$\Delta^{2,4,6,8,10}$-Tetradec	Ac	H	*E. tirucalli*	[130]
366	Hexadec	Ac	H	*Stil. loranthacea*	[159]
367	H	Ac	(9Z)-Octadec	*C. tiglium*	[89]
368	H	Ac	(9Z,12Z)-Octadec	*C. tiglium*	[89]
369	*i*Bu	*i*Bu	Ac	*E. nubica*	[140]
370	Bz	*i*Bu	Ac	*E. nubica*	[140]
371	Dec	Ac	Dec	*C. alienus*	[193]

C. = Croton; E. = Euphorbia; Pe = Pedilanthus; Sa. = Sapium; Stil. = Stillingia

Table 21 Hydroxylated 4α-4-deoxy phorbol esters (possibly isolation artefacts)

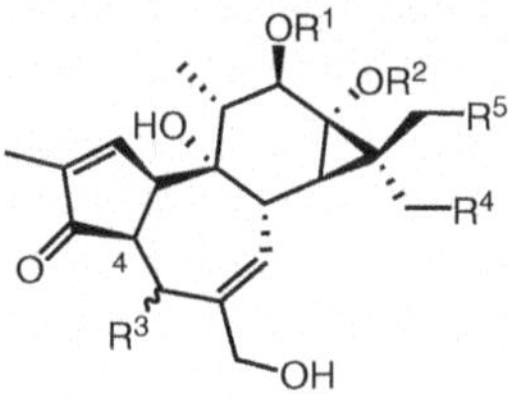

Compound	R¹	R²	R³	R⁴	R⁵	Source	References
372	Me-2Abz-OH	Ac	β-OH	H	H	*Sa. indicum*	[137]
373	2,3MeBu	*i*Bu	H	H	OH	*E. aellenii*	[189]
374	Hex	Ac	H	OH	H	*Sa. insigne*	[109]
375	Dodec	Ac	H	OH	H	*Sa. insigne*	[109]
376	Hexadec	Ac	H	OH	H	*A. fordii* *A. fordii*	[113] [114]
377	Hex	Ac	H	OAc	H	*Sa. insigne*	[109]

A = Aleurites; E. = Euphorbia; Sa. = Sapium

tiglianes of the 5β-hydroxy-6,7-α-epoxy series do not undergo retroaldol epimerization (see Sect. 4.2), suggesting that epoxidation of the ring B double bond on the α-face destabilizes the 4α-configuration, presumably for steric reasons.

Table 22 4α-4,12-Dideoxy phorbol esters (possibly isolation artefacts)

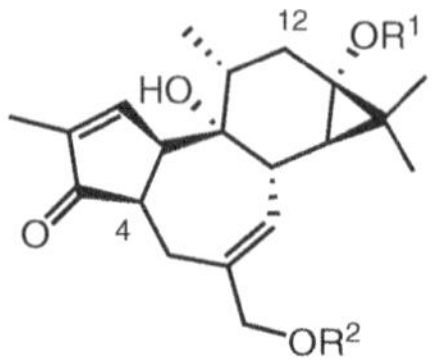

Compound	R^1	R^2	Source	References
378	$\Delta^{3,5}$-Tetradec	H	*Ex. bicolor*	[177]
379	Hexadec	H	*E. guyoniana*	[194]
380	Ac	Ac	*Ex. bicolor*	[177]
381	2,3MeBu	Ac	*E. pithyusa*	[176]

E. = Euphorbia; Ex = Excoecaria

Deoxygenated 4α-4-Deoxyphorbol Esters

Both mono- and bis-deoxygenated tiglianes are included in this group, with the sites of deoxygenation closely related to those of the corresponding and native 4β-series (**378–388**; Tables 22 and 23). The frequency of 5β-hydroxylation associated with deoxygenation of the allylic 20 position (Table 23) is remarkable, since so few tiglianes from the 4β-series are hydroxylated at C-5, and this could be associated with a higher affinity of the enzyme responsible for the hydroxylation of C-5 for tiglianes deoxygenated at C-20, when compared to those hydroxylated at this position.

7-Hydroxy(Hydroperoxy, Oxo) Δ^5-Phorbol Esters

These compounds (**389–438**, Table 24) are the largest class of probable artefacts in the tigliane inventory, vividly testifying to the sensitivity of phorbol esters to auto-oxidation (see Sect. 4.1). Auto-oxidation generates, with an ene reaction-type mechanism, compounds from the Δ^5-7-hydroperoxy series, by undergoing a formal dismutative process that generates compounds from the corresponding 8-hydroxy and 8-oxo series. Only compounds resulting from oxygen attack from the convex β-face occur, which is in accordance with the results of photooxygenation studies on phorbol esters (see Sect. 4.2).

20-Dehydro Phorbol Esters and Carboxylates

From the auto-oxidation of phorbol, in addition to 7-hydroperoxy-Δ^5-tiglianes, also the 20-dehydro derivatives **439–450** are formed (see Sect. 4.2). Due to the electron-withdrawing properties of the formyl group, the Δ^6-double bond remains therefore unscathed. The 20-demethyl-20-carboxyl derivative **450** could result from overoxidation of the corresponding 20-dehydro derivative **446**, and both compounds have been isolated from the same plant extract (Table 25).

Table 23 4α-4,20-Dideoxy phorbol esters (possibly isolation artefacts)

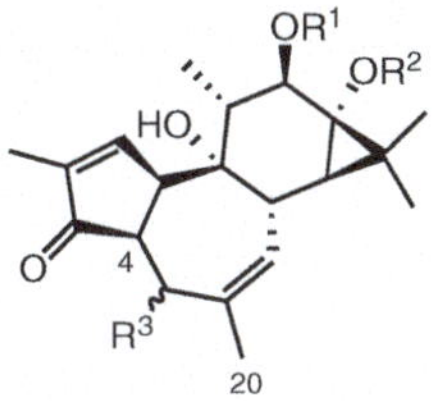

Compound	R¹	R²	R³	Source	References
382	Ac	*i*Bu	β-OH	*E. prolifera*	[182]
383	Me-2Abz-OH	Ac	β-OH	*Sa. indicum*	[109, 137]
384	Bz	*i*Bu	β-OH	*E. neorubella* *E. dracunculoides* *E. dendroides* *E. nematocypha*	[174] [181] [195] [196]
385	Bz	(*S*)-2MeBu	β-OH	*E. neorubella*	[174]
386	Ac	*i*Bu	H	*E. clementei*	[197]
387	Me-2ABz-OH	Ac	H	*Sa. indicum*	[109, 137]
388	Bz	*i*Bu	H	*E. bupleiroides* *E. sanctae-catharinae*	[198] [180]

E. = *Euphorbia; Sa.* = *Sapium*

Seco-Tiglianes

Cleavage of the exocyclic cyclopropane bond at the C/D-ring junction of tiglianes generates two major classes of phorboids, the daphnanes (C-14/C-15 cleavage) and the rhamnopholanes (C-13/C-15 cleavage) (Fig. 7). Compounds from these classes of D-*seco*-phorboids will be covered in a forthcoming separate contribution in this book series, and we will discuss here only B-ring *seco*-tiglianes, the only additional class of naturally occurring *seco*-tiglianes.

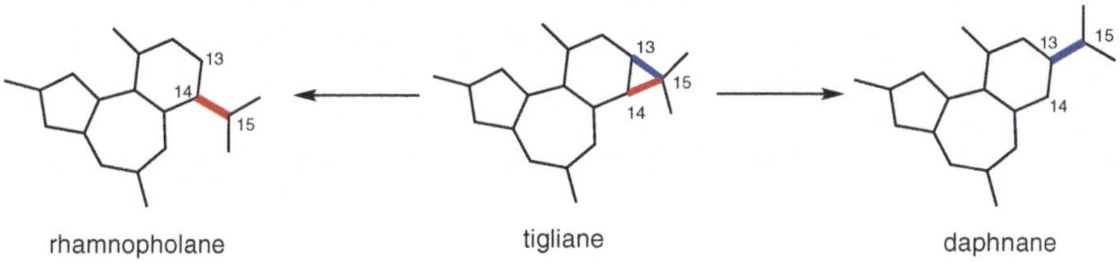

Fig. 7 Cyclopropane ring-cleavage of the tigliane skeleton to generate the daphnane and the rhamnopholane skeleta

Table 24 Δ^5-phorbol esters (possibly isolation artefacts)

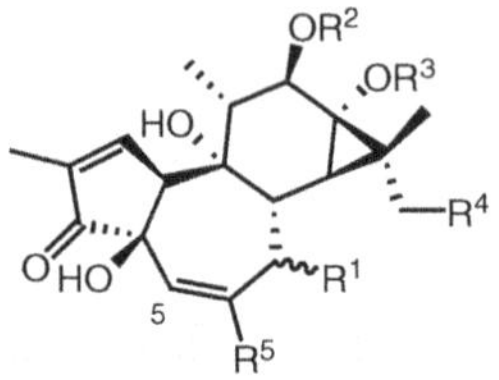

Compound	R^1, R^2	R^3	R^4	R^5	R^6	Source	References
389	β-OH, α-H	H	Dodec	H	CH_2OH	*R. trisperma*	[160]
390	β-OH, α-H	OH	*i*Bu	H	CH_2OH	*C. tiglium*	[191]
391	β-OH, α-H	OH	2MeBu	H	CH_2OH	*C. tiglium*	[191]
392	β-OH, α-H	OAc	2MeBu	H	CH_2OH	*C. tiglium*	[199]
393	β-OH, α-H	OTigl	*i*Bu	H	CH_2OH	*C. tiglium*	[92]
394	β-OH, α-H	OTigl	2MeBu	H	CH_2OH	*C. tiglium*	[92]
395	β-OH, α-H	OBz	Oct	H	CH_2OH	*W. lamatsoensis*	[99]
396	β-OH, α-H	OBz	Oct	H	CH_2OH	*W. scytophylla*	[123]
397	β-OH, α-H	H	*i*Bu	OAng	CH_2OH	*E. grandicornis*	[200]
398	=O	H	Ac	H	CH_2OH	*E. fischeriana* *P. prostrata*	[144] [145]
399	=O	H	Oct	H	CH_2OH	*W. lamatsoensis*	[99]
400	=O	H	Dodec	H	CH_2OH	*R. trisperma*	[160]
401	=O	H	Tetradec	H	CH_2OH	*R. trisperma*	[160]
402	=O	H	Hexadec	H	CH_2OH	*R. trisperma*	[160]
403	=O	OH	Ac	H	CH_2OH	*C. tiglium*	[201]
404	=O	OH	2MeBu	H	CH_2OH	*C. tiglium*	[191]
405	=O	OH	Dec	H	CH_2OH	*C. tiglium*	[92]
406	=O	OAc	iBu	H	CH_2OH	*C. tiglium*	[199]
407	=O	OAc	2MeBu	H	CH_2OH	*C. tiglium*	[199]
408	=O	O(*S*)-2MeBu	Ac	H	CH_2OH	*C. damayeshu*	[202]
409	=O	O(*S*)-2MeBu	*i*Bu	H	CH_2OH	*C. damayeshu*	[202]
410	=O	O(*S*)-2MeBu	(*S*)-2MeBu	H	CH_2OH	*C. damayeshu*	[202]
411	=O	O(*S*)-2MeBu	Oct	H	CH_2OH	*C. damayeshu*	[202]
412	=O	O(*S*)-2MeBu	Dec	H	CH_2OH	*C. damayeshu*	[202]
413	=O	O(*S*)-2MeBu	(4*Z*)-Dec	H	CH_2OH	*C. damayeshu*	[202]
414	=O	O(*S*)-2MeBu	(4*Z*,7*Z*)-Dec	H	CH_2OH	*C. damayeshu*	[202]
415	=O	OTigl	*i*Bu	H	CH_2OH	*C. tiglium*	[92, 191]

(continued)

Table 24 (continued)

Compound	R^1, R^2	R^3	R^4	R^5	R^6	Source	References
416	=O	OTigl	Prop	H	CH_2OH	*C. tiglium*	[92]
417	=O	OTigl	Oct	H	CH_2OH	*C. damayeshu*	[202]
418	=O	OTigl	Dec	H	CH_2OH	*C. damayeshu*	[202]
419	=O	OTigl	(4*Z*)-Dec	H	CH_2OH	*C. damayeshu*	[202]
420	=O	OBz	Oct	H	CH_2OH	*D. holosericea* *W. scytophylla* *W. lamatsoensis*	[203] [123] [99]
421	=O	OHex	Oct	H	CH_2OH	*W. scytophylla*	[123]
422	=O	ODec	Ac	H	CH_2OH	*D. genkwa*	[104, 204]
423	=O	O(2*E*,4*E*)-Dec	Ac	H	CH_2OH	*D. genkwa*	[104, 105]
424	=O	ODodec	Ac	H	CH_2OH	*T. howii*	[205]
425	=O	OHexadec	Ac	H	CH_2OH	*Al. fordii* *Sa. sebiferum*	[113] [109]
426	=O	OHexadec	Ac	OH	CH_2OH	*Al. fordii* *Al. fordii*	[113] [114]
427	=O	H	*i*Bu	OAng	CH_2OH	*E. grandicornis*	[200]
428	=O	OTigl	*i*Bu	H	CHO	*C. tiglium*	[92]
429	β–OOH, α–H	H	Ac	H	CH_2OH	*P. prostrata*	[145]
430	β–OOH, α–H	H	Oct	H	CH_2OH	*W. lamatsoensis*	[71]
431	β–OOH, α–H	OTigl	*i*Bu	H	CH_2OH	*C. tiglium*	[92]
432	β–OOH, α–H	OTigl	2MeBu	H	CH_2OH	*C. tiglium*	[92]
433	β–OOH, α–H	OBz	Oct	H	CH_2OH	*W. scytophylla* *W. lamatsoensis* *St. chamaejasme*	[123] [99] [98]
434	β–OOH, α–H	OBz	Dec	H	CH_2OH	*St. chamaejasme*	[98]
435	β–OOH, α–H	ODec	Ac	H	CH_2OH	*C. mauritianus*	[106]
436	β–OOH, α–H	O($\Delta^{2,4}$)-Undec	Ac	H	CH_2OH	*O. paniculata*	[110]
437	β–OOH, α–H	OHexadec	Ac	H	CH_2OH	*Sa. sebiferum*	[109]
438	O-	O2MeBu	Hex	O2MeBu	CH_2O-	*C. caudatus*	[125]

E. = Euphorbia; R. = Reuteralis; O = Ostodes; P. = Pimelea; Sa. = Sapium; St. = Stellera; T. = Thymelaea; W. = Wikstroemia

Ring B-*Seco*-Koglianes and Their Rearrangement Products

Opening of ring B triggers some of the rare skeletal tigliane rearrangements of biogenetic relevance. Despite a dense functionalization and the inherent strain of the tetracyclic tigliane ring system, Nature seems to have largely overlooked the possibility of further elaborating the tigliane connectivity, leaving it to the excitement of chemists, as discussed in Sect. 4.1. Thus, only two rearranged tiglianes have been

Table 25 20-Dehydro phorbol esters and 20-carboxylates (possibly isolation artefacts)

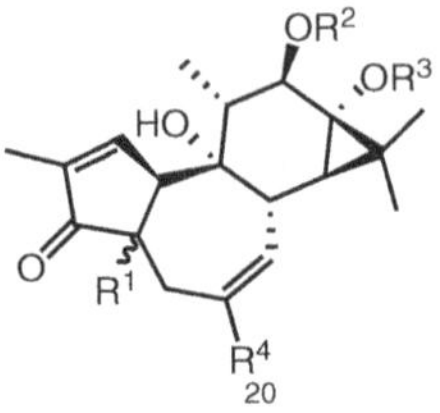

Compound	R^1	R^2	R^3	R^4	Source	References
439	α-H	OH	*i*Bu	CHO	*C. damayeshu*	[187]
440	α-H	OH	(*S*)-2MeBu	CHO	*C. damayeshu*	[187]
441	α-H	OTigl	(*S*)-2MeBu	CHO	*C. damayeshu*	[187]
442	α-H	OTigl	Ac	CHO	*C. tiglium*	[191]
443	α-H	O(Me-2Abz-OH)	Ac	CHO	*Sa. indicum*	[109, 115]
444	α-H	OHexadec	Ac	CHO	*Stil. loranthacea*	[159]
445	β-H	O(Me-2Abz-OH)	Ac	CHO	*Sa. indicum*	[109, 115, 194]
446	β-OH	H	Ac	CHO	*E. fischeriana*	[80, 145]
447	β-OH	H	Dodec	CHO	*Stil. loranthacea* *R. trisperma*	[159] [160]
448	β-OH	H	(5*Z*)-Tetradec	CHO	*Stil. loranthacea*	[159]
449	β-OH	H	Hexadec	CHO	*E. fischeriana* *E. ebracteolata*	[82, 116, 145, 165] [206]
450	β-OH	H	Ac	COOH	*E. fischeriana*	[80]

E. = *Euphorbia*; *R.* = *Reuteralis*; *Sa.* = *Sapium*; *Stil.* = *Stillingia*; *W.* = *Wikstroemia*

reported as natural products, in stark contrast to the over 400 diterpenoids with the canonic tigliane skeleton.

Pedrolide

The diester pedrolide (**451**) was isolated from *Euphorbia pedroi* Molero and Rovira, a plant endemic to Portugal, in the course of studies aimed at discovering reversal of P-glycoprotein-associated multidrug resistance [207]. Its structure was established by a combination of spectroscopic and X-ray crystallographic studies, although the poor resolution of the crystallographic data did not allow confirmation of the relative and absolute configurations. Pedrolide (**451**) was suggested to be derived from the 4α-20-deoxy tigliane diester **388** [180, 198] via an intramolecular Michael reaction leading to the C-6 cation **452**, next evolving into the 1,7-cyclo 4(5 → 6)*abeo* intermediate **453** by extrusion of C-5 and water trapping of the C-5 primary cation.

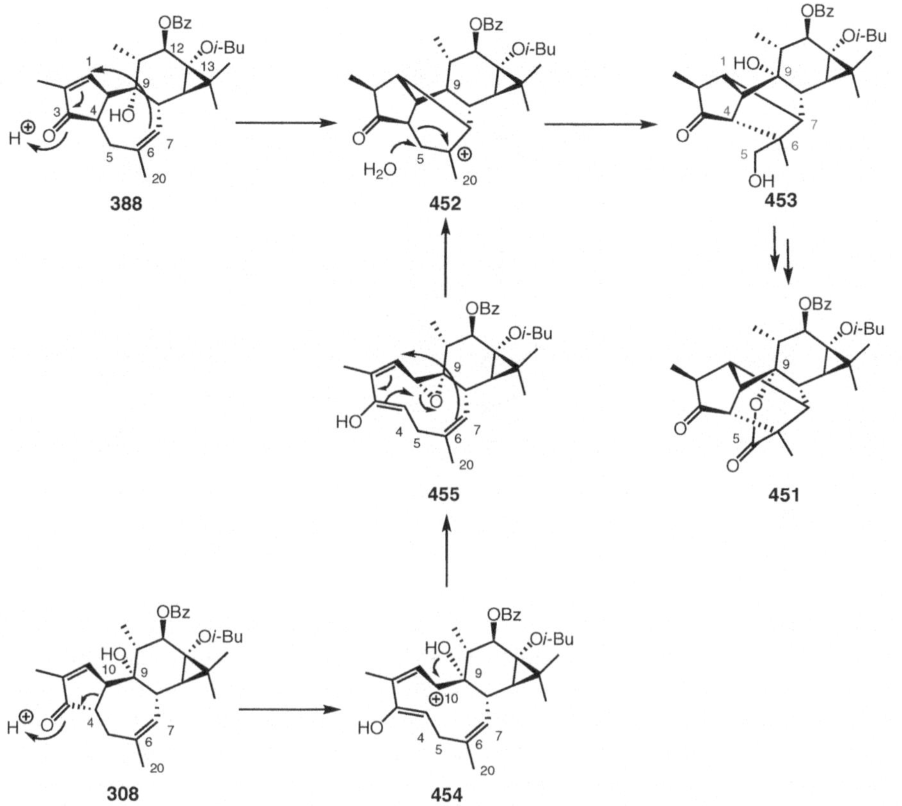

Scheme 2 Possible formation of pedrolide (**451**) from the tigliane diesters **308** and **388**

Oxidation of the C-5 hydroxy group to a carboxyl and then lactonization with the 9-hydroxy group eventually affords **451** (Scheme 2). However, given the artefact status of 4α-phorbol derivatives, an alternative formation of the C-6 cation **452** from the naturally occurring 4β-4-deoxytigliane diester **308** [139, 180] would also be worthy of consideration. Thus, cleavage of the C-4/C-10 bond could provide via the *seco*-cation **454**, and then via trapping by the 9-hydroxy group (possibly concerted with the cleavage) to the alternative macrocyclic epoxide substrate **455** and the formation of **452** (Scheme 2).

Euphordraculoate A

The rearranged tigliane euphordraculoate A (**456**) was isolated from *Euphorbia dracunculoides* Lam., a plant distributed all over the Northern Hemisphere subtropical biome, from Spain to the Arabian Peninsula, Western India and South Central China [208]. Euphordracunculoate A (**456**) could be derived from the retroaldol isomerization of the co-occurring 5-hydroxy-20-deoxytigliane precursor **310** [174,

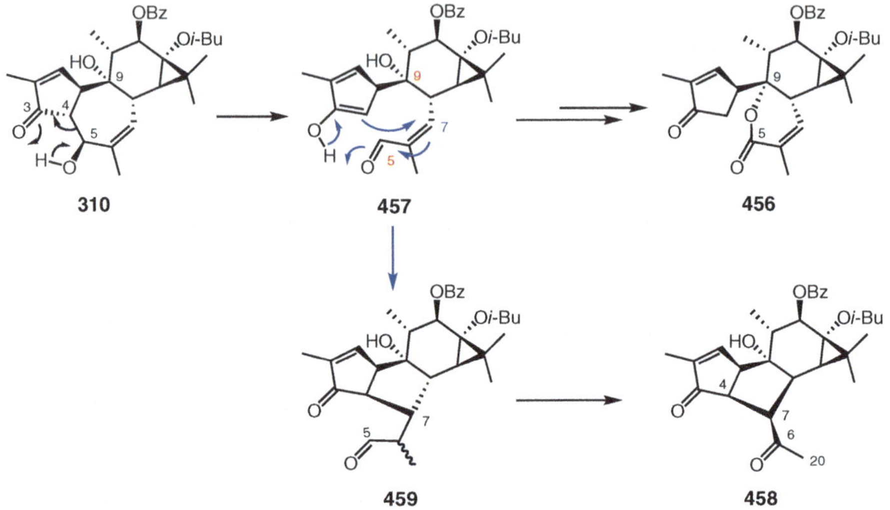

Scheme 3 Possible formation of euphordraculoate A (**456**) and euphordraculoate B (**458**) from the 5-hydroxy-20-deoxytigliane diester **310**

179, 181, 182] to the B-*seco*-aldehyde **457**, followed by hemiacetal formation between the 5-aldehyde and the 10-hydroxy group and oxidation to the corresponding lactone (Scheme 3). Attempts to mimic this process by treatment of precursor **310** with bases (NaH) unleashed a distinct rearrangement, leading to a compound unhappily named, just like the natural product **456**, euphordracunculoate B (**458**) despite the presence of a lactone ring and not of a carboxylate group. Rather than 1,2-attack from the 9-hydroxy group, the B-*seco* C-5 enal **457** underwent 1,4-Michael addition of the ring A dienolate to the enal generated by the opening of the B-ring. The α,α-disubstituted aldehyde **459** resulting from Michael addition then underwent oxidative degradation, presumably via dioxetane formation, to eventually afford the 5-*nor*-4(5 → 1)abeotigliane **458**. This compound, but neither its precursor **310** nor the natural product **456**, could inhibit the Wnt/β-catenin signaling pathway, thereby qualifying as a potential anticancer agent [208].

3.4 Biogenetic Considerations

Kant famously remarked that, because of the absence of natural laws, chemistry "should not be regarded as science, but nothing more than systematic art" [209]. The natural law of secondary metabolites is their biosynthesis, and, in this context, tiglianes enjoy a Kantian state, since the biosynthesis of their skeleton is mechanistically enigmatic, still undefined from an enzymatic standpoint, and confused by analogy considerations [210, 211]. Thus, the carbon connectivity of tiglianes is

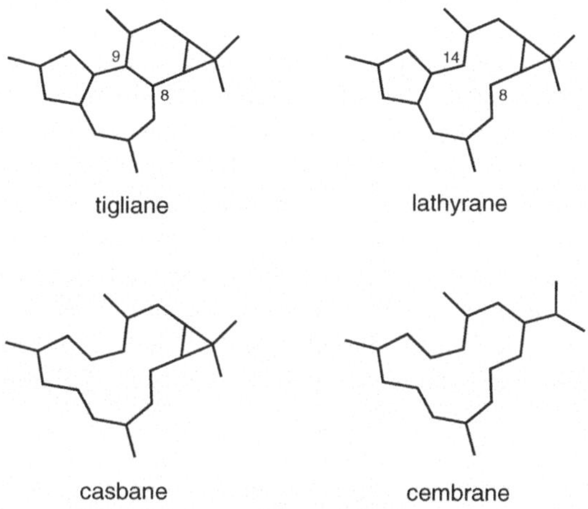

Fig. 8 Topological relationship between the lathyrane and the tigliane diterpenoid skeleta and their relationship with those of cembranes and casbanes

related topologically to that of the lathyranes, with the only difference being the presence of an extra bond between C-8 and C-9 (C-8 and C-14 in lathyrane numbering; Fig. 8). This similarity has supported the widespread view of a direct relationship between the two classes of compounds, witheliminating tiglianes originating from the transannular cyclization of lathyrane precursors, and lathyranes firmly entrenched into the head-to-tail branch of diterpenoid biosynthesis, stemming, via a bicyclic casbane intermediate, from a monocyclic cembrane precursor (Fig. 8) [212].

This view of the lathyrane-tigliane biogenetic relationship is, however, at odds with the presence of a distinct functionalization pattern in compounds from the two classes with the most remarkable difference being the hydroxylation of one of the ring-junction cyclopropane carbons, a feature lacking in lathyranes but present in all tiglianes (see below for the discussion of a hapax). In addition, a lathyrane-tigliane direct biogenetic relationship would involve the enzymatic hydroxylation of a cyclopropane carbon, a reaction facing severe mechanistic challenges in terms of enzymology and chemical plausibility. Taken together, these topologic and mechanistic observations suggest a more complex scenario.

Comparison of the functionalization profile of lathyrol (**460**) [213] and phorbol (**1**), the "founding fathers" of, respectively, the lathyrane and tigliane diterpenoid families, shows puzzling differences. The methyl substitution along the carbon sequence of lathyranes has a C_2 axis passing through the methyl-bearing cyclopentane carbon (C-2 in the lathyrane numbering) and the cyclopropane carbon bearing the *gem*-dimethyls (C-10 in the lathyrane numbering, Fig. 8). This symmetry is mirrored in tiglianes by a symmetrical location of the methyl substituents around the same cyclopentane-cyclopropane axis (C-2—C-15 in the tigliane numbering, Fig. 8). This analogy suggests a straightforward biogenetic relationship between the two classes of diterpenoids, interconverted by the formation/cleavage of a bond between C-8 and C-9 (tigliane numbering). On the other hand, Nature seems to have been deviously more subtle.

Thus, if the two classes of compounds were related by a simple transannular cyclization, only the functionalization around the cyclization sites should differ significantly, while that around the rest of the ring system, and especially at the other terminus, should be similar. However, this is not the case. Thus, the peripheral carbon sequence of lathyrol (**460**) around the methyl-bearing cyclopentane carbon shows oxygenation at the β- and γ-carbons when read clockwise (red arrow), and at the α′- and γ′-carbons when read anti-clockwise (blue arrow). Conversely, phorbol (**1**) is only oxygenated at the γ- carbon for the corresponding clockwise sequence, and at the α′- and β′-carbons for the anti-clockwise sequence, with an (*E*)-configured δ-double bond in the clockwise sequence of lathyrol (**460**) and a (*Z*)-configured δ′-double bond in the opposite sequence of phorbol (**1**).

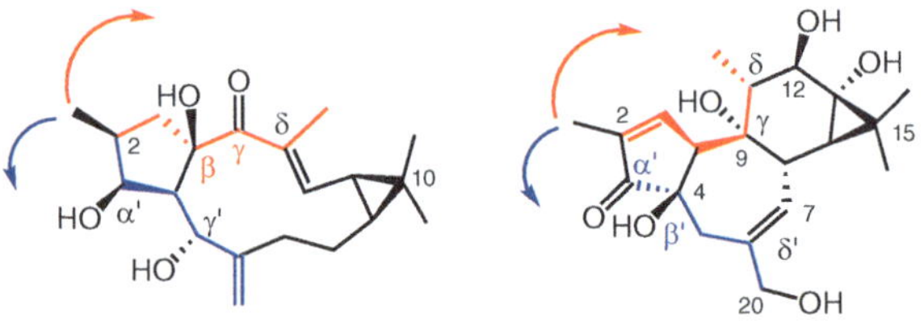

460 (lathyrol) **1** (phorbol)

If the analysis of the functionalization profile is extended to the whole carbon skeleton and to additional members of the two diterpenoid families [214], three further critical observations emerge:

(a) All tigliane derivatives have *α*-oriented tertiary oxygen functions at C-13 and C-9, that is, at the C/D- and the B/C-ring junctions, suggesting a biogenetic relationship between the formation of these junctions and their oxygenation. Only one tigliane derivative oxygenated at C-9 and lacking hydroxylation at C-13 like the ingenanes has been reported [215]. This compound (**461**) is a veritable phytochemical hapax also for its lack of ester decoration, an unprecedented feature within phorboids, and was isolated from *Euphorbia hylonoma* Hand.-Mazz., a plant used in traditional Chinese medicine as a purgative [215]. Given its trace concentration and the lengthy fractionation required for its purification, **461** could be an artefact. Nevertheless, its structure raises interesting biogenetic issues, independently of its status of native (enzyme-derived), or a programmed (reactivity-derived) natural product. *Euphorbia hylonoma* contains a complex mixture of ingenanes and ingenane-derived diterpenoids and only a single tigliane constituent. It is therefore not unconceivable that **461** could derive from the acid-triggered rearrangement of the corresponding ingenane derivative **462** (Scheme 4), a reaction documented under laboratory conditions [216] and supported by the presence of other diterpenes associated with the acidic cyclization of ingenane precursors [215]. On the othere hand, the presence of the 9,10-epoxide hints also at a connection with lathyranes, where oxygenation at C-14 and C-15 (corresponding to C-9 and C-10 in tigliane numbering) is very common, and for which the cyclopropane ring is never hydroxylated.

Scheme 4 Possible derivation of the C-13 deoxygenated tigliane **461** from the ingenane precursor **462**

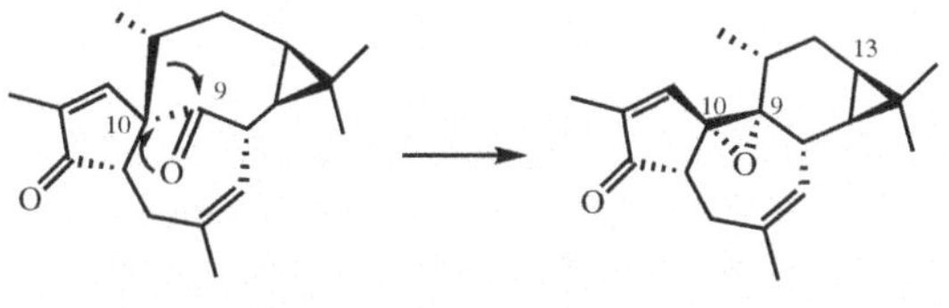

(b) Additional hydroxylation of the parent cyclic framework of phorbol is limited to C-5, while in lathyranes and related compounds nuclear oxygenation can take place at almost every carbon with the remarkable exception of this part of the cyclopropane ring (C-9 and C-11) and of C-1 [214, 217].

(c) No lathyrane derivative with a complementary functionalization at the two carbons involved in the putative transannular cyclization to tiglianes is known. In naturally occurring lathyranes, C-14 is generally a carbonyl and therefore electrophilic, while C-8 is either hydroxylated or part of an unfunctionalized sequence of methylenes, lacking therefore the nucleophilic character associated with the presence of a double bond or induced by an adjacent carbonyl. In synthetic model compounds, evidence for the possibility of this transformation has been documented, obtaining, however, compounds for which their similarity with natural tiglianes does not go over the connectivity of the diterpenoid core [218]. Nature seems therefore to have avoided this chemically plausible transannular reaction-based strategy, relying instead on a more complex domino process, capable of associating ring closure with functionalization at the cyclopropane carbon.

When taken together, these topological considerations would be better accommodated by an orthogonal rather than a sequential biogenetic relationship between lathyranes and tiglianes.

An additional issue for a possible lathyrane-tigliane biogenetic relationship emerges from the need to hydroxylate the cyclopropane carbon of lathyranes either before or after the C-4—C-14 transannular cyclization. Several lines of evidence make this reaction questionable under both laboratory and biological conditions. Thus, oxygen atoms are introduced in isoprenoid scaffolds by specific cytochrome P450 monooxygenases acting with an oxygen-rebound mechanism that involves homolysis of C-H bonds and the generation of radicals [219]. This chemistry, as well as that involving a cyclopropyl cation, would be unsuitable for the functionalization of a cyclopropane carbon. Thus, the cyclopropane C-H bonds are significantly stronger than the C-H bonds of alkanes (444 vs. 423 kJ/mol of bond energy) [220], and radical and cation formation would therefore be directed on the adjacent carbons. As a matter of fact, cyclopropanation is a popular medicinal chemistry maneuver to reduce the susceptibility of a carbon to oxidative metabolization by P450 enzymes [220]. Enzymes can overcome the selectivity rules of test-tube wet chemistry maneuver, but the cyclopropyl radical is highly unstable and prone to fragmentation [221], while the cyclopropyl cation rapidly springs open to an allyl cation through a disrotatory electrocyclic process [222]. These mechanistic considerations cast serious doubt on the plausibility of a direct enzymatic hydroxylation of a cyclopropane carbon. Thermodynamic and kinetic considerations therefore require that naturally occurring heteroatom-bearing cyclopropanes should be generated by mechanism(s) basically different from those in which heteroatoms are introduced in alkyl chains or rings, with formation of the carbon-heteroatom bond preceding formation of the three-membered ring. An analysis of the occurrence of the heteroatom-substituted cyclopropane motif in natural products fully supports this view.

The presence of a cyclopropane ring is not uncommon in natural products, and especially in isoprenoids, with corner- and edge-protonated cyclopropopyl cations being important intermediates in the rearrangements of carbenium ions en route to cyclic isoprenoid skeletons [223, 224]. Conversely, the heteroatom-substituted cyclopropane motif is very rare in natural products, and mainly documented in the isosteric amino-substituted version rather than in the hydroxylated version. The most important heteroatom-functionalized cyclopropane natural product is 1-aminocyclopropane carboxylic acid (**463**) [225], the precursor of the plant hormone ethylene [226], but additional aminocyclopropanes have been isolated from bacteria, as exemplified by colibactin (**464**) [227], genotoxic toxins associated with colon cancer [227], and the 2,3-methano amino acids coronamic acid (**465**) [228, 229] and allocoronamic acid (**466**) [230].

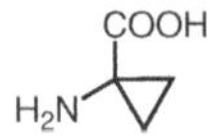

463 (aminocyclopropane carboxylic acid)

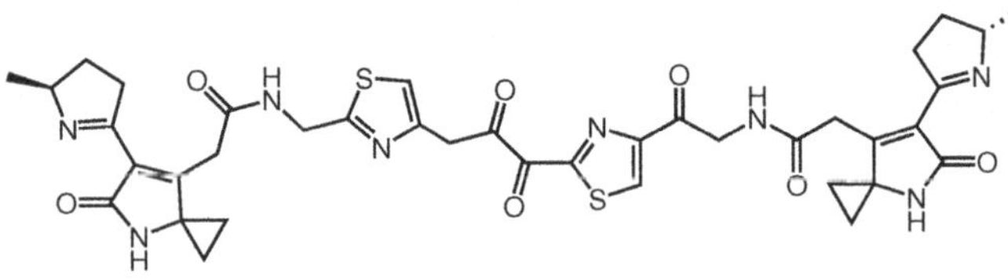

464 (colibactin)

465 (coronamic acid)

466 (allocoronamic acid)

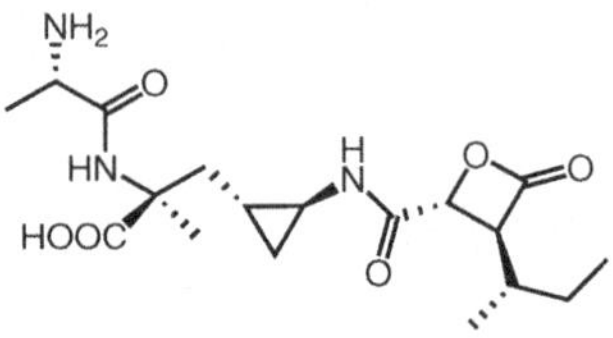

468 (belactosin A)

In these compounds, the cyclopropane ring is generated in a nitrogen-containing precursor by a cyclase enzyme that uses a carbanion-based intramolecular displacement reaction (Scheme 5) [231]. Mechanistic diversity stems from the nature of the enolate stabilizing group (thioester for **465** and **466** [228], pyridoxine for **463** [228]) and that of the leaving group (chloride for **465**, **466** [228], and thiomethyl adenosine for **463** [228]), with nitrogen functionalization preceding, in all cases, the closure of the cyclopropane ring. A strategically similar process operates in the formation

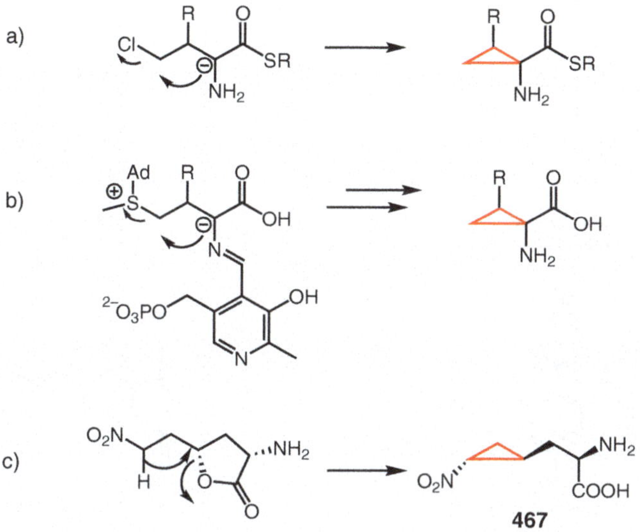

467

Scheme 5 Formation of the amino cyclopropane motif by a carbanion-based intramolecular displacement reaction (Ad = adenosin-5′-yl)

of the two residues of 3-(2-nitrocyclopropyl)alanine (**467**) contained in the depsipeptide antibiotic hormaomycin [232], with closure of the cyclopropane ring taking advantage of the acidifying properties of the nitro group (Scheme 5, c). A nitrocyclopropane intermediate has also been suggested to be involved in the formation of the aminocyclopropane moiety of belactosin A (**468**) [233].

Tiglianes aside, very few natural products with a hydroxy(alkoxy)cyclopropane structure are known, all of which are of microbial origin except the fern glucosides **469**, **470** [234], hortiolide C (**471**)-related rutaceous limonoids [235], and the furanochromanone membranaceumone A (**472**) [236]. The microbial inventory includes the polyketide pairs malleicyprol (**473**)/bis-malleicyprol (**474**) [237], toblerols A and H ((**475**) and (**476**), respectively) [238], the amino acid cleonine (**477**) [239, 240], its methyl derivative diplonine (**478**) [241], and the CD45 tyrosine phosphatase inhibitor penitricin D (**479**) [242]. The involvement of a cyclopropanol derivative in the biosynthesis of the meroterpenoid talaromyolide A (**480**) has been questioned [243], while coprine (**481**), a fungal alcohol dehydrogenase inhibitor, is actually a cyclopropanone hemiaminal [244], and will not be discussed herein.

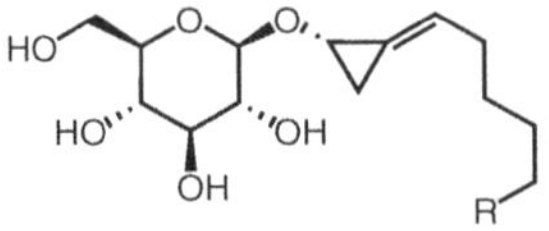

R
469 CH_2OH
470 COOH

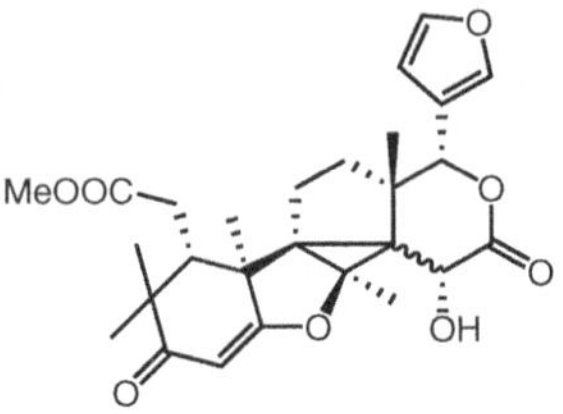

471 (hortiolide C)

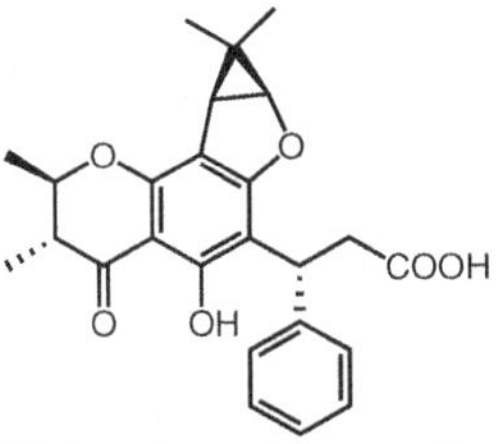

472 (membranaceumone A)

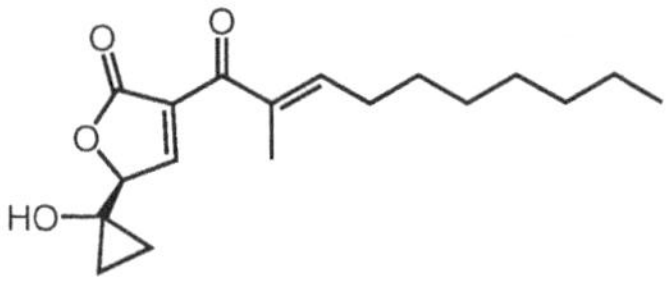

473 (maleicyprol)

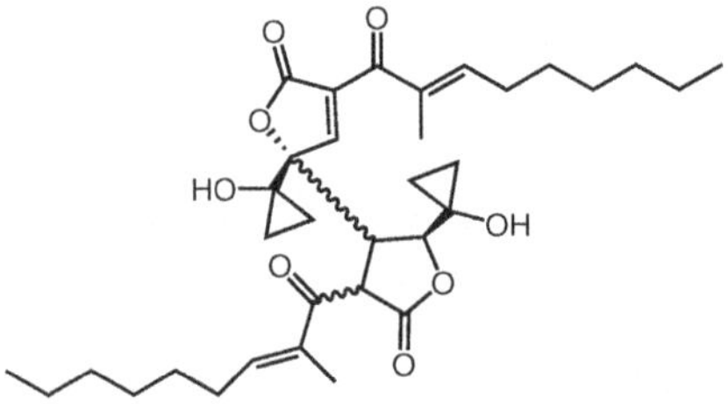

474 (bis-maleicyprol)

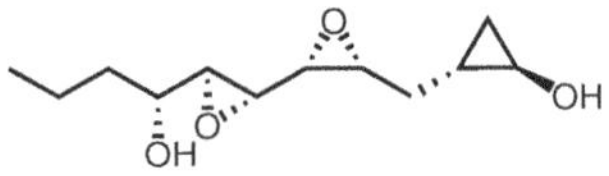

475 (toblerol A)

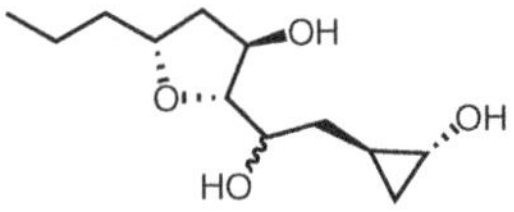

476 (toblerol H)

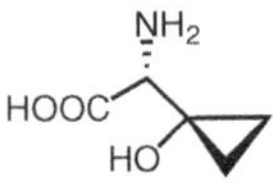

477 (cleonine)

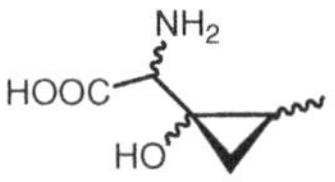

478 (diplonine)

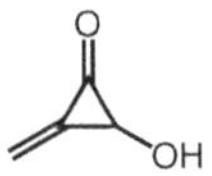

479 (penitricin D)

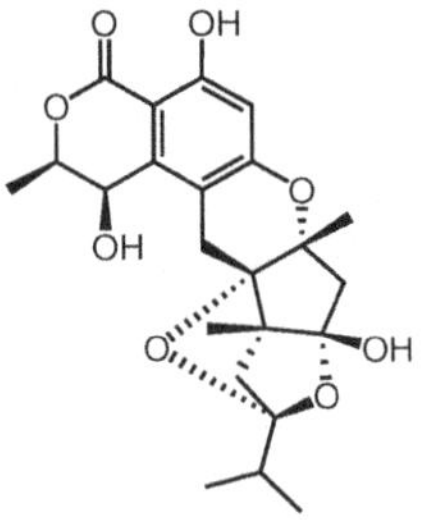

480 (talaromyeline A)

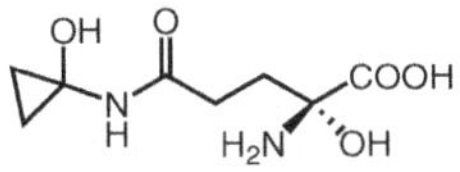

481 (coprine)

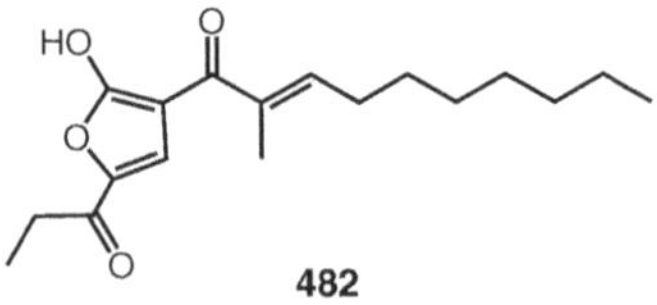

482

In malleicyprol (**473**) and bis-malleicyprol (**474**) the cyclopropanol moiety acts as a warhead. Upon single-electron oxidation-triggered ring opening, it generates a resonance-stabilized β-keto radical, which eventually evolves into a propanoyl group (**482**) [237]. It has also been proposed that the cyclopropanol moiety of tiglianes can generate species capable of covalently binding macromolecules, but this theory was disproved on the basis of biological as well as chemical evidence, as discussed in Sect. 6. As a matter of fact, the cyclopropanol moiety is "armed" only in **473** and **474**, being "disarmed" in all other natural cyclopropanols.

Malleicyprols (**473, 474**) are the only cyclopropanols for which their biosynthesis has been elucidated enzymatically [237]. This structural element is generated in a five-carbon acidic precursor bearing an ω-dimethylsulfonium group and an α,β-diol system (Scheme 6). NAD^+-mediated dehydrogenation generates an ene-diol system, which displaces in a S_N2-fashion dimethylsulfide from the ω-carbon and generates a cyclopropanol oxonium ion that is next reduced by NADH. Overall, the α-carbon undergoes an unusual NAD^+-NADH-mediated transient redox cycle, a sort of hydride-rebound mechanism, with the carboxylate and the oxygen functions on the α-and β-carbons involved in a bonding network with two magnesium ions [237] (Scheme 6).

The hortiolide C (**471**) cyclopropanol moiety has been suggested to be derived from a furanoid precursor **483** via an intramolecular Michael reaction involving the furanyl β-carbon as nucleophile and the electrophilic vinyl substituent at the furanyl α-carbon producing **484** as an intermediate [235] (Scheme 7).

Cleonine (**477**) is a constituent of the macrolide anticancer antibiotic cleomycin B2 [239] and of the depsipeptide valgamycin C [240]. Cleomycin B2 was first

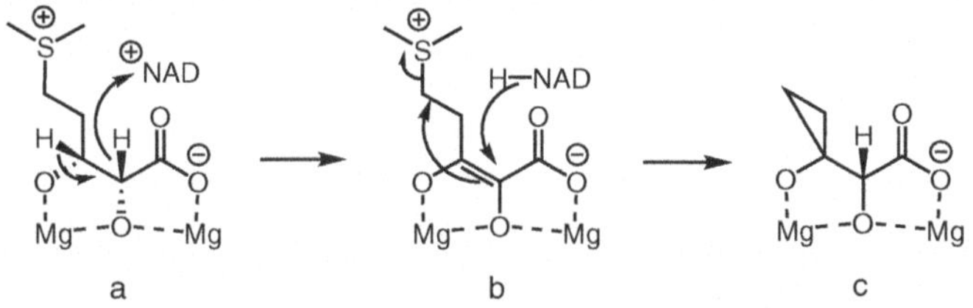

Scheme 6 Formation of the cyclopropanol moiety of malleicyprols **473** and **474**

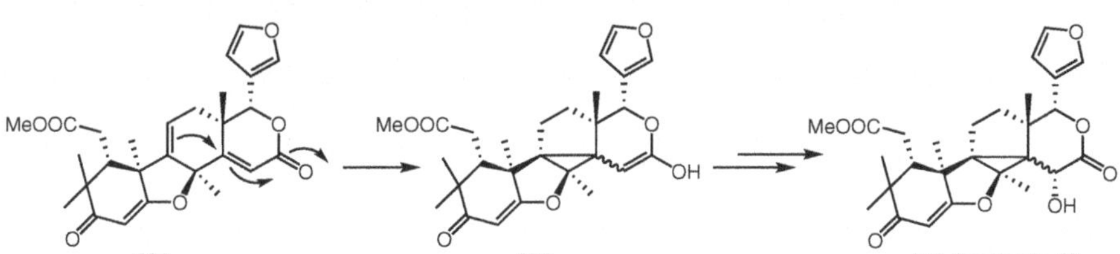

483 **484** **471** (hortiolide C)

Scheme 7 Putative formation of the cyclopropanol moiety of hortiolide C (**471**) from the furanoid precursor **483** via the enol **484**

obtained from a UV-irradiated strain of a bleomycin-producing actinomycete [239], but was also then isolated, as a constituent of the antibiotic valgamycin C, from a non-mutated microorganism [240]. No bioactivity has been associated with cleonine (**477**), but its *C*-methyl derivative diplonine is a neurotoxin produced by the ear-rot fungus (*Stenocarpella maydis* (Berk.) Sutton = *Diplodia maydis* (Berk.) Sacc.), a pathogen of maize that causes diploidiosis, a degenerative neurological condition of ruminants [241]. The configuration of **477** and diplonine has been controversial. Cleonine (**477**) was first suggested to have an (*S*)-configuration by analogy with the (*S*)-configuration of β-hydroxyvaline, an amino acid occurring in Actinomycetes and assumed to be its precursor [239]. However, Marfey analysis of cleonine from valgamycin C showed a (*R*)-configuration, casting doubt on the biogenetic relationship with hydroxyvaline [240]. The cyclopropyl moiety of toblerols (**475**, **476**) [238] has been suggested to originate from the oxidative cleavage of a cyclopropane carbon-carbon bond associated with the fragmentation of a larger precursor, and a similar process could be involved in the formation of the cyclopropanol glucosides **469** and **470** [234]. Finally, the cyclopropyl ether of the furanochromanone membranaceumone A (**472**) [236] has been suggested to be derived from the chromanyl ether **485** by quenching of a tertiary cation by β-proton loss (Scheme 8), a common process in isoprenoid cationic chemistry.

In all these cases, the scenario for the formation of the cyclopropanol moiety is quite different from the context where this element needs to be generated in phorbol, since the polyhydroxycarbonyl milieu necessary for a carbanion-based displacement process is difficult to accommodate in a pre-tigliane skeleton, while a fragmentative oxidative origin would imply a higher isoprenoid precursor. Despite our limited knowledge, it is nevertheless remarkable that, in all cases where the biosynthesis of heteroatom-substituted cyclopropanes was elucidated, the heteroatom was not introduced after the formation of the cyclopropane ring, but was already present in the precursor.

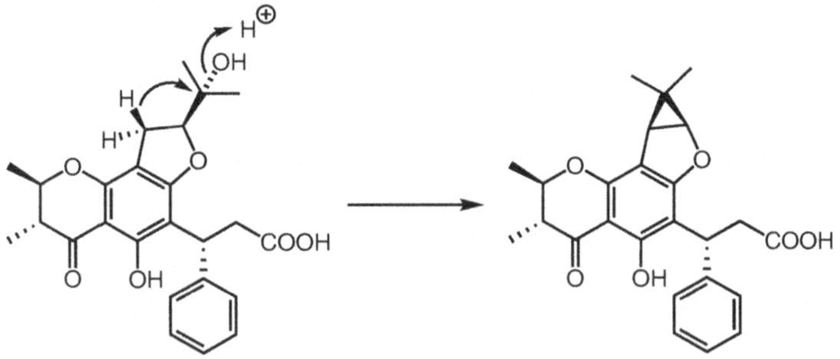

485 **472** (membranaceumone A)

Scheme 8 The formal 1,3-elimination leading to closure of the cyclopropyl ring of membranaceumone A (**472**) from the isoprenylated chromone precursor **485**

Two decades ago Jasmin Jakupovic (1949–2005), then at the Technische Universität Berlin, proposed an alternative to the lathyrane-tigliane relationship that brilliantly solved the critical issues associated to this relationship. Jasmin Jakupovic passed away before he could publish his ideas, which remained confined in the PhD thesis of his collaborator Franck Jesche, and in private discussions with one of the authors of this review (G.A.) and other colleagues. Based on this material, and with the help of the Jakupovic family (his wife Fotini and his son Sven), a detailed summary of his view was presented in the ingenane chapter of volume 102 of this series [245]. We will now expand the most critical section of this proposal, namely, that associated with the formation of the cyclopropanol moiety. This structural motif can be easily generated readily under laboratory conditions, but with reactions (the titanium-based Kulincovich reaction, or the carbene addition to an enol ether [246]) lacking an equivalent within the mechanistic portfolio of enzymatic reactions.

Jakupovic proposed that the tigliane and the lathyrane skeleta could originate from casbenes having a different configuration around the Δ^3-double bond, the one adjacent to the cyclopropane ring (Scheme 9, **B** and **B′**, respectively), or, alternatively, from geranylgeranyl or geranyllinalyl pyrophosphates. A transannular cyclization between the homoallyl and the allyl carbons of the central double bond would directly generate the (*E*,*E*)-lathyrane (6,10-cyclocasbane) structural motif **C′** from a Δ^3-(*E*)-casbene precursor **B′**. The same process from the Δ^3-(*Z*) casbene precursor **B** would generate the alternative 6,10-(*Z*,*E*)-cyclocasbene **C**. This, after introduction of a carbonyl group at the methylene adjacent to the cyclopropane ring, would directly generate the tigliane skeleton in a possibly concerted process involving (the tigliane numbering is used throughout).

(a) protonation of the 13-carbonyl of **D**;
(b) a cyclopropylcarbonyl-cyclopropylcarbinyl rearrangement (cyclopropane "walk") via anionic migration of the cyclopropane bonds and generation of the allyl cation at C-8;
(c) transannular cyclization involving quenching of the incipient C-8 cation by the C-11 double bond with generation of a cationic center at C-11 (yielding **E**);
(d) hydride migration from C-9 and generation of a cationic center at C-10 resulting in **F**, and
(e) quenching of the positive charge at C-10 by a water molecule to provide **G**.

The presence in miglianes of an identical stereochemical signature around all carbons involved in these steps suggests that these events might occur in a concerted, domino-type way inside a single enzyme, by analogy to the "black box" of terpene cyclases. Additional steps of hydride- and carbon shifts would then be involved in the generation of the ingenane skeleton [245].

Jakupovic's proposal brilliantly solves the critical issues involved in the lathyrane-tigliane biogenetic relationship, which is orthogonal and not sequential, with the cyclopropane "walk" causing the frameshift mutation responsible for the distinct functionalization profile of compounds from the two classes. The derivation of tiglianes and lathyranes from distinct casbene stereoisomers via distinct processes

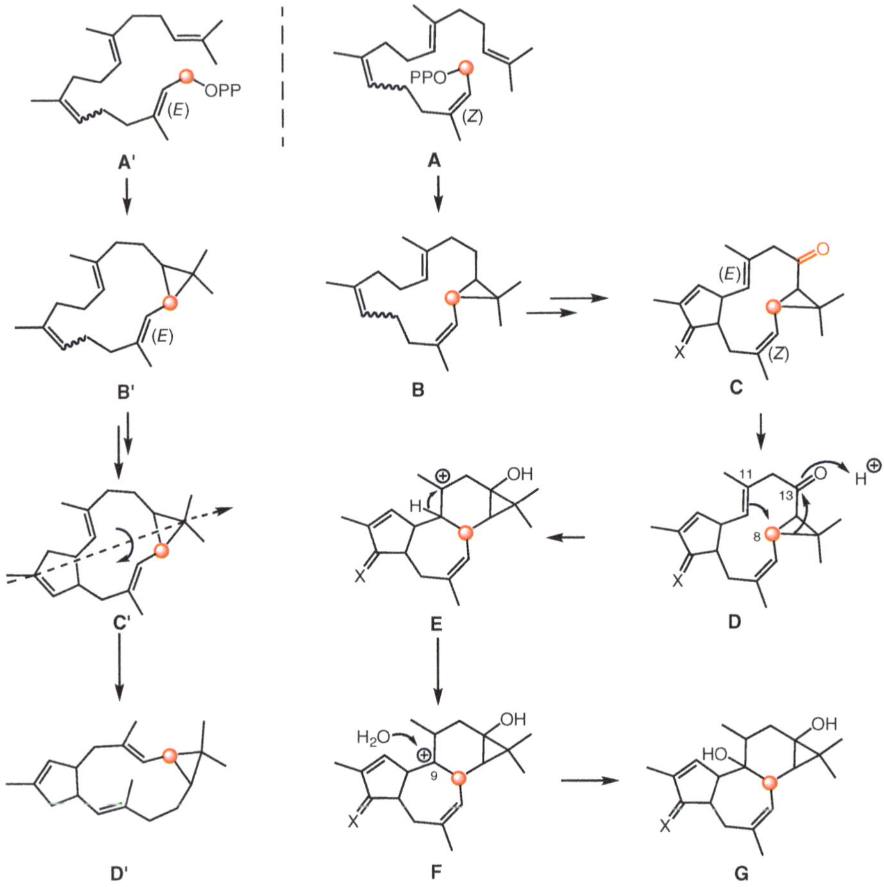

Scheme 9 The Jakupovic hypothesis for the biosynthesis of tiglianes. The C-1 carbon of geranylgeranyl pyrophosphate is indicated in order to highight its different location in lathyranes (C′) and in tiglianes (F) in relation to the cyclopropane moiety. For the sake of clarity, the configurations at stereogenic carbons have been omitted

is supported by the observation that compounds from the two classes have never been found to co-occur in the same plant. Lathyranes have, however, been observed to co-occur with ingenanes, where the cyclopropane ring is, with very few exceptions, not hydroxylated [245]. This suggests that a cyclopropane "walk" triggered by protonation of a C-13 hydroxy group has minor biogenetic dominance than that generated by protonation of a C-13 carbonyl group.

The Jakupovic proposal has, undoubtedly, the effect of lighting a lamp in a dark chamber, and is worth serious investigation with suitably labeled precursors. Thus, tracing the location of the cyclopropane *gem*-dimethyl substituted carbon would be an obvious litmus test to prove the *abeo*-nature of the cyclopropane ring in tiglianes when compared to casbanes and lathyranes, and could be done using suitably labeled geranylgeranyl precursors.

Carbonyls have a low-lying LUMO and cyclopropane bonds a high-lying HOMO due to their increased *p*-character compared to mainstream Csp^3–Csp^3 bonds, engaging with "enthusiasm" into Wagner-Meerwein rearrangements, as exemplified by the phorbol-phorbobutanone rearrangement (see Sect. 4.3). Nevertheless, the cyclopropane "walk" involved in the Jakupovic proposal of tigliane biosynthesis does not seem to have been observed in cyclopropyl ketones under laboratory conditions. It could be noted, however, that the unpromising behavior of these compounds in [3+2] cycloaddition reactions has limited their interest to the organic chemistry community [247]. On the other hand, a mechanism similar to that proposed by Jakupovic has been postulated to rationalize the long enigmatic rearrangement of the monoterpene chrysanthenone (**486**) to filifolone (**488**) via the cation **487** (Scheme 10), while additional examples of a cyclopropane "walk" associated with carbocation rearrangements have been documented or postulated in the realm of isoprenoid chemistry and biosynthesis.

In the chrysanthenone to filifolone rearrangement, migration of one cyclobutane bond to a carbonyl and formation of a hydroxylated cyclopropylmethyl cation **487** are terminated by ring expansion and generation of a cyclobutanone [248], with the driving force being the sequential generation of an allyl cation and then of a stable carbon-oxygen double bond (Scheme 7). In the realm of the non-carbonyl based versions of this process, a cyclopropane "walk" of was first proposed by Dauben to rationalize the ring expansion of the sesquiterpene thujopsene (**489**) to widdrol (**491**) via the cation **490** (Scheme 11), and next proved by labeling experiments [249]. Incidentally, it was Dauben who in 1962 reported the structure elucidation of cembrene, the "father" of all head-to-tail diterpenoids [250].

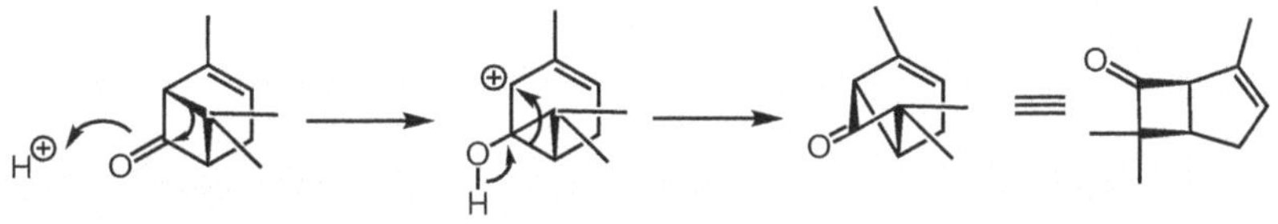

486 (chrysanthenone) **487** **488** (filifolone)

Scheme 10 The chrysanthenone (**486**)–filifolone (**488**) rearrangement

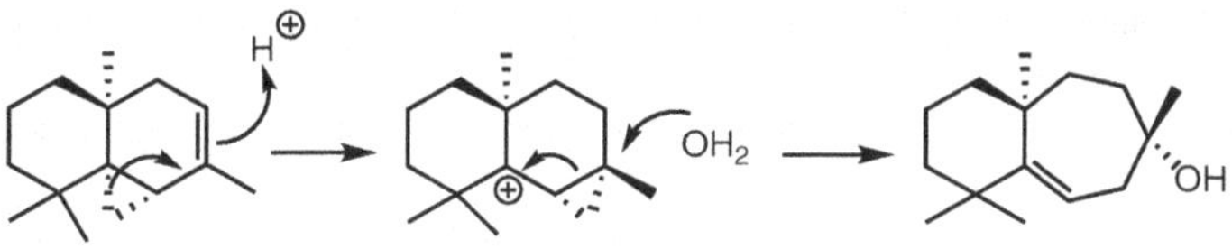

489 (thujopsene) **490** **491** (widdrol)

Scheme 11 Ring expansion of the tricyclic sesquiterpene thujopsene (**489**) to widdrol (**491**)

A similar cyclopropylmethyl-cyclopropylmethyl cation rearrangement was discovered to occur in the biosynthesis of cyclooctatin **492** (Scheme 12) [251]. The diterpene skeleton of this compound, an inhibitor of lysophospholipase isolated from the bacterium *Streptomyces melanosporofaciens* MI614-43F2, is similar to the sesterpenoid skeleton of the ophiobilins, and it was suggested to originate from a similar mechanism, with direct generation of the cyclooctane ring in a polyolefin cyclization reaction [252]. However, labeling experiments showed a different scenario, with the terpene cyclase CotB2 generating the cyclopropane-substituted central seven-membered ring intermediate **494** from **493**. This then undergoes a cyclopropane "walk" via a 1,3-anionotropic migration of the cyclopropane bond distal to the cationic center, followed by ring expansion by cleavage of one cyclopropane bond and the eventual formation of **492** (Scheme 12) [251].

A cyclopropane "walk" was also postulated to rationalize the formation of the diterpene *neo*-verrucosan-5β-ol (**496**) from the tetracyclic cation **495** (Scheme 13) in the thermophilic bacterium *Chloroflexus aurantiacus*, and the process was analyzed in detail with the aid of DFT calculations [253].

Taken together, topologic and mechanistic considerations provide a solid rationale to investigate the Jakupovic proposal, also including the generation of different casbene stereoisomers, and the way a putative 6,10-cyclocasbene tigliane precursor is formed. The formation of isomeric casbene derivatives could simply involve geranylgeranyl- and geranyllinalyl phosphate precursors, a well-established biogenetic process to generate (*E*)/(*Z*) bonds in cyclic isoprenoid frameworks. As an alternative, distinct conformations of a geranyllinalyl phosphate precursor could

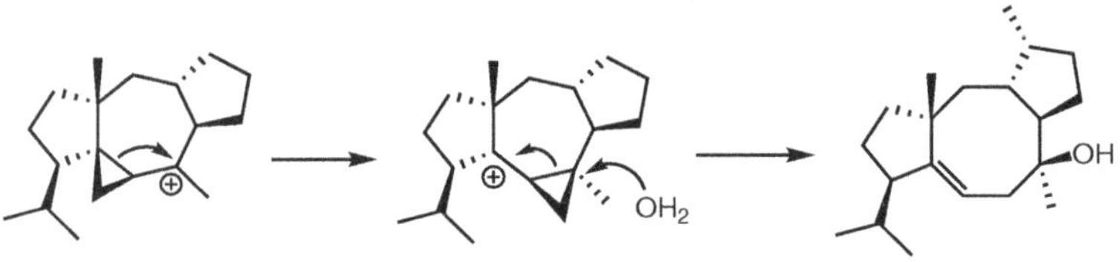

493 **494** **492** (cyclooctatin)

Scheme 12 The cyclobutane "walk" involved in the biosynthesis of the diterpene cyclooctatin (**492**)

Scheme 13 The cyclopropane "walk" involved in the proposed biosynthesis of the diterpene *neo*-verrucosan-5β-ol (**496**)

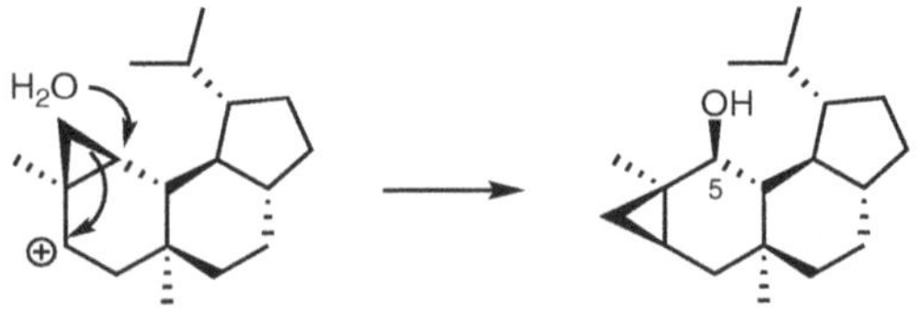

495 **496** (*neo*-verrucosan-5β-ol)

be involved, an observation that could explain the occurrence of the *cis*- and *trans*-fusion of the cyclopropane ring in some lathyrane derivatives [254]. The closure of the five-membered ring might also be associated with functionalization, since the five-membered ring of tiglianes and lathyranes bears multiple oxygenations. Thus, the dioxocasbene (**497**) has been suggested to cyclize spontaneously, after isomerization of the allylic alcohol system to a carbonyl, to the lathyrane jolkinol C (**498**) via an intramolecular aldol reaction [212]. Additional support for this viewpoint is needed, but, given the almost ubiquitous presence of oxygenation at C-3 and C-12 in lathyranes, the mechanism could be general for this class of compounds. On the other hand, a distinct mechanism should operate in tiglianes, since the dioxygenated carbons on ring A are adjacent, while, in some cases, ring A is monooxygenated.

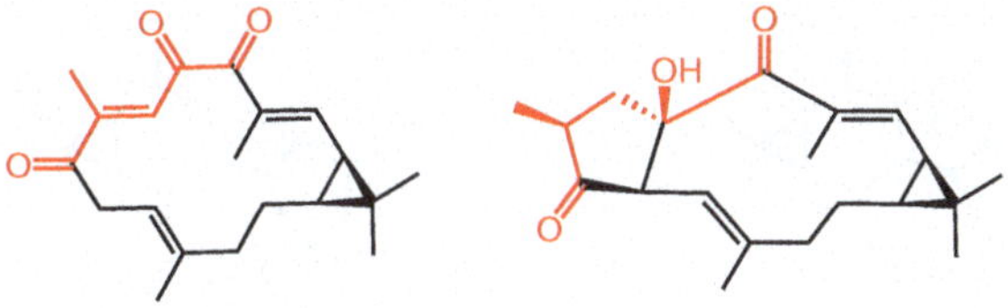

497 **498** (jolkinol C)

The tigliane skeleton has a central role in the biogenesis of phorboids. Its relationship with the ingenanes has been discussed extensively in Volume 102 of this book series [245], while that with the daphnanes and rhamnopholanes will be included in a forthcoming review on diterpenoids from this class.

In conclusion, despite the major importance of phorbol (**1**) in the chemical and biological literature [255], its biosynthesis has so far received scarce attention. When critically analyzed, the lathyrane-tigliane relationship via a simple transannular cyclization would require an unprecedented enzymatic performance in the generation of the cyclopropanol moiety and a profligacy of distinct oxidase enzymes to differentially functionalize compounds of the two classes. The Jakupovic proposal brings back the tigliane biosynthesis to mainstream enzymatic chemistry, postulating a complex transannular cyclization which, in a concerted way, generates the tigliane skeleton and introduces oxygen functions at the cyclopropane carbon C-13 and at C-10. Being the result of organic chemistry considerations only, it will need biochemical studies to proceed from the status of a mechanistic Occam's razor to that of an accepted pathway, but its confirmation or rejection will, undoubtedly, provide the opportunity to further advance our knowledge on the way isoprenoids are produced by plants.

4 Reactivity

Within the major tigliane polyols, only phorbol (**1**) can be obtained from the hydrolysis/transesterification of its corresponding naturally occurring acyl derivatives in amounts sufficient to sustain systematic reactivity studies, since the other polyols are either unstable under the conditions of hydrolysis/transesterification (4-deoxyphorbol (**499**), 16-hydroxyphorbol (**12**), and 5β-hydroxy-6α-epoxyphorbol (**11**)) or, although widespread, are not accumulated to comparable and significant amounts (12-deoxyphorbol (**13**)). Most reactivity data of tigliane derivatives therefore have been obtained on phorbol (**1**), and thus its chemistry dominates this Section.

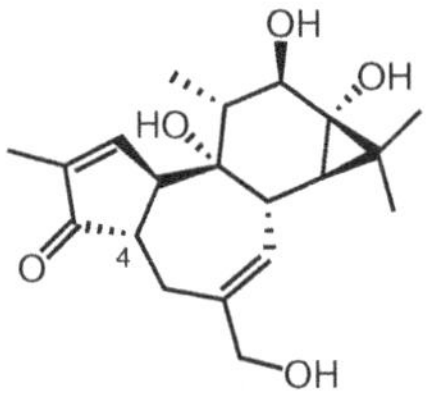

499 (4-deoxyphorbol)

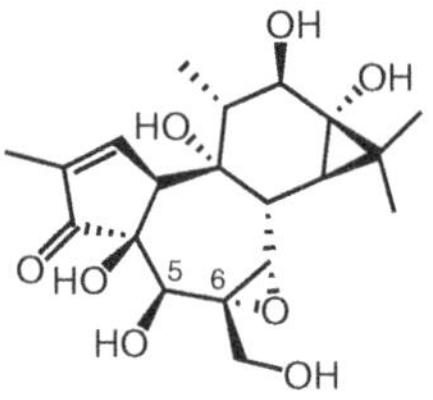

11 (5β-hydroxy-6α-epoxyphorbol)

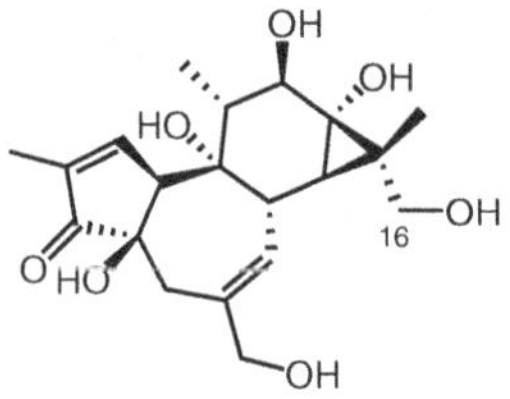

12 (16-hydroxyphorbol)

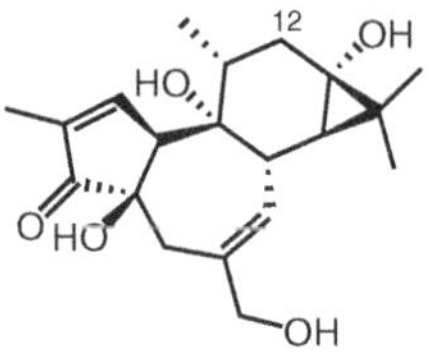

13 (12-deoxyphorbol)

The tetracyclic carbon skeleton of phorbol (**1**) is decorated with five hydroxy groups, a ketone carbonyl and two double bonds, one of which is conjugated to the carbonyl (Fig. 9). The tertiary hydroxy group at C-9 is generally solitary in terms of reactivity, while the other functionalities combine in three hotspots of reactivity: the ring A α,β-unsaturated tertiary ketol system spanning C-1 to C-4, the ring B allylic alcohol motif involving C-6, C-7, and C-20, and the glycol system involving the C/D ring junction carbon (C-13) and its adjacent ring C oxymethine (C-12). Owing to the relative insulation of these functional motifs, "explosive" rearrangements leaving little structural evidence of the original connectivity and typical of highly functionalized diterpenoids like taxoids [256] have never been reported in tigliane derivatives. A reductive, hotspot-based analysis of the reactivity of **1** is therefore possible, and this section is organized accordingly. Additional functionalities, like the hydroxy groups at C-5, C-9, and C-16, can bridge the reactivity of the three reactivity islands, and will be analyzed independently.

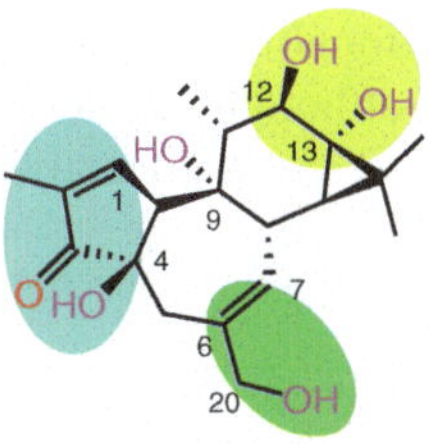

1 (phorbol)

Fig. 9 Reactivity of phorbol

4.1 The Ring A α,β-Unsaturated Tertiary Ketol System

The C-1–C-2 Double Bond

The ring A carbonyl-conjugated double bond of phorbol (**1**) shows little tendency to give addition reactions with nucleophiles, an observation disproving the early proposal that phorbol esters act in vivo as thia-Michael acceptors, irreversibly binding to biomolecules [257]. It is also less reactive than the $\Delta^{6,7}$ double bond toward hydrogenation (see Sect. 4.1), while the lack of reactivity with electrophilic reagents is an obvious consequence of the conjugation with the C-3 carbonyl. This profile of poor reactivity towards nucleophiles could in part be associated with the incomplete planarity between the carbonyl and the double bond, evident from the low ε value of the UV absorption at 230 nm (ca. 5,000 M^{-1} cm^{-1} in ethanol) [52] (see Sect. 3.2). In phorbol derivatives from the 4α-series, photochemical excitation of the enone chromophore by irradiation with a low-pressure mercury lamp (λ ca. 254 nm) triggers an intramolecular [2πs+2πs] cycloaddition with the $\Delta^{6,7}$-double bond, producing the cage-like skeleton of lumiphorbol (**500**) [258], for which the structure was solved by X-ray crystallographic analysis of lumiphorbol 12,13,20-triacetate (**501**) [258]. The facile photochemical cycloaddition is favored by the cup-shape of the A/B-ring system of 4α-phorbol (**14**), which positions the two double bonds on roughly parallel planes, optimizing the interaction of their orbital systems.

The conjugated $\Delta^{1,2}$-double bond was reactive in a formal [2+1] cycloaddition with the stabilized sulfur ylide generated from the alkynyl amide-containing sulfonium salt **502** [259]. The initial Michael addition on phorbol 12,13-dibutyrate (**503**), a semi-synthetic ester, occurred from the α-face of the double bond, biasing the subsequent sulfur displacement step, and leading to the cyclopropane derivative **504**. No selectivity was observed in the generation of the third cyclopropane stereocenter, and the reaction gave in modest yield (24%) a 3:2 mixture of two diastereomers of **9** [259]. Despite the basic conditions of the reaction, formation of the cyclopropane adduct was not accompanied by epimerization at C-4, a clear indication that the initial thia-Michael addition was faster than the vinylogous retro-aldol epimerization (see Sect. 4.1). Thia-Michael reactions on synthetic nor-tigliane enones were also reported, but with this functionality in a non-natural location [260]. Interestingly, some of these compounds, although devoid of interaction with protein kinase C, showed cancer cell line-selective cytotoxic activity [260].

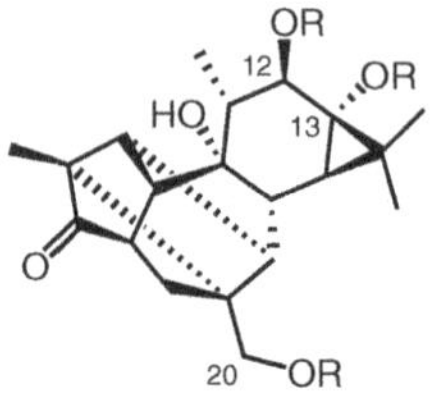

	R
500	H (lumiphorbol)
501	Ac

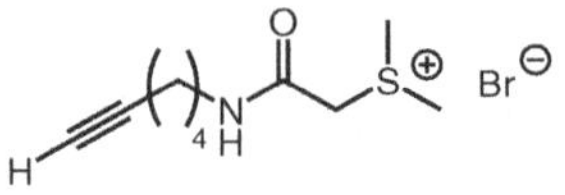

502

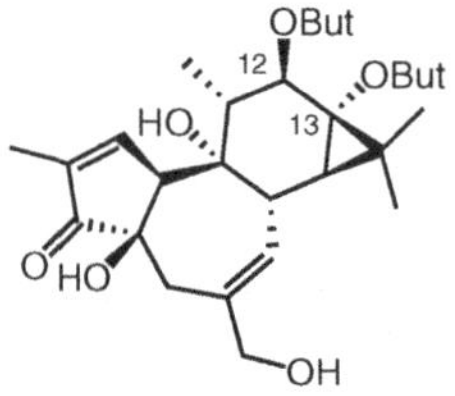

503 (phorbol 12,13-dibutyrate)

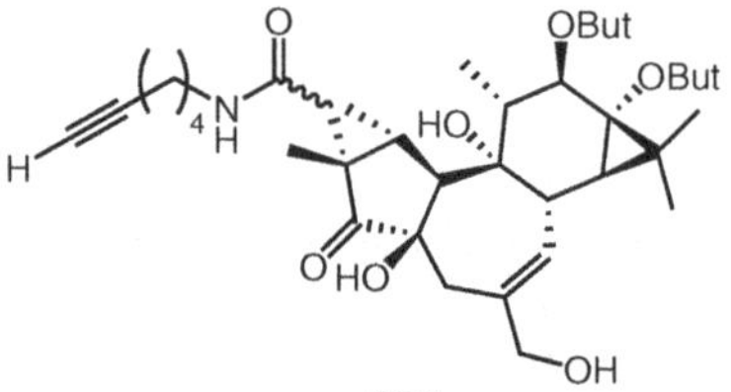

504

The C-3 Ketone Carbonyl

The C-3 ketone carbonyl group is unreactive with nitrogen nucleophiles (hydrazine, hydroxylamine) used in pre-spectroscopic times to detect the presence of this functional group, and was therefore considered cryptic in terms of analytical reactivity. A small yield of a 2,4-dinitrophenylhydrazone could only be obtained under forced conditions, but not under the standard conditions of the test [261]. Since models do not show any significant steric hindrance, the reason of this lack of reactivity is unclear: the incomplete conjugation with the adjacent double bond should moderate the decrease of electrophilicity associated with conjugation, while the inductive effect and the hydrogen-bonding capacity of the adjacent hydroxy group at C-4 are actually expected to increase reactivity with nucleophiles. The 3-keto group can be reduced with hydrides [261]. The stereoselectivity was reported to be surprisingly higher with $LiAlH_4$, which exclusively afforded the 3β-alcohol (3β-dihydrophorbol (**505**)) resulting from attack from the less hindered α-face, while $NaBH_4$ gave a mixture of epimeric alcohols, with prevalence of the 3β-diastereomer [261]. Exclusive formation of the 3β-alcohol was observed with the Luche reagent ($NaBH_4$ and $CeCl_3$) [261, 262]. The 3α-alcohol (3α-dihydrophorbol (**506**)) could be obtained stereoselectively under conditions of chelation-controlled reduction, using tetrabutylammonium triacetoxyborohydride [262, 263]. Under these conditions, the reaction takes place intramolecularly, only after the activation for hydrogen transfer provided by the exchange of one of the electron-withdrawing acetate ligands with the 4-hydroxy group. In borohydride reactions, ester groups were generally retained at the C-20 and the C-13 hydroxy groups, the most prone to hydrolysis, but when the reaction was carried out on phorbol pentaacetate (**507**), migration of acetyl from the tertiary 4-acetate to the 3β-hydroxy group took place [261]. The 3β-dihydro derivative of phorbol 12,13,20-triacetate (**508**) was cleaved with sodium periodate in dioxane-water to the unstable ring A-*seco*-ketoaldehyde **509**, characterized as its borohydride reduction product **510** [264]. These ring-cleavage reactions are now only of historical relevance, but were important prior to the availability of spectroscopic methods to gain information on the functional groups present in phorbol (**1**).

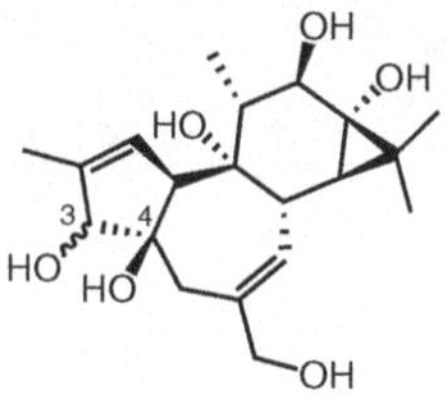

505: 3β-OH
506: 3α-OH

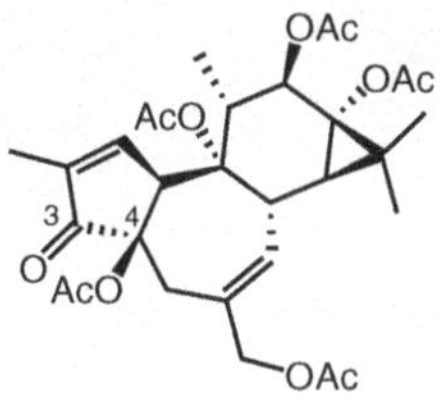

507

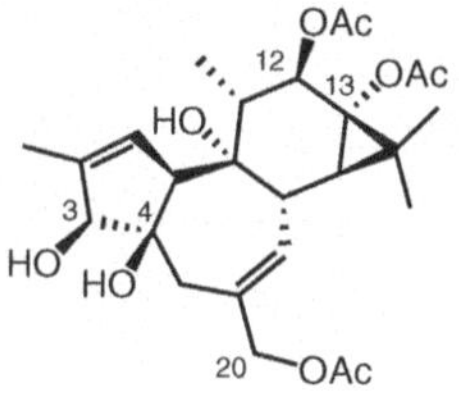

508

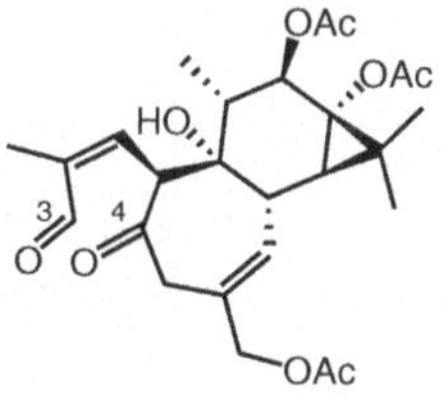

509

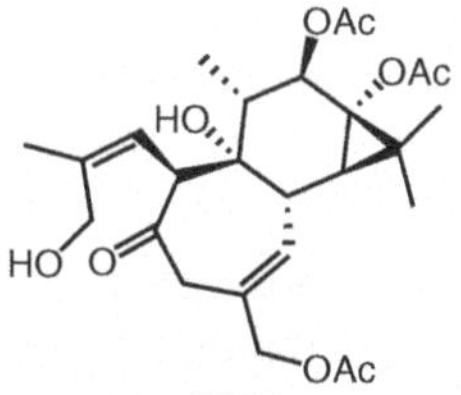

510

The C-4 Carbon Atom

In most natural tiglianes, C-4 is hydroxylated as in phorbol (**1**), but analogues from the 4-deoxy series also occur, generally in lower amounts when compared to their C-4 hydroxylated analogues (see Sect. 3), in a variety of euphorbiaceous plants (see Sect. 3.3). Both compounds from the 4-hydroxy- and the 4-deoxy series are unstable in basic conditions, with epimerization of the substituent at C-4 (hydroxy

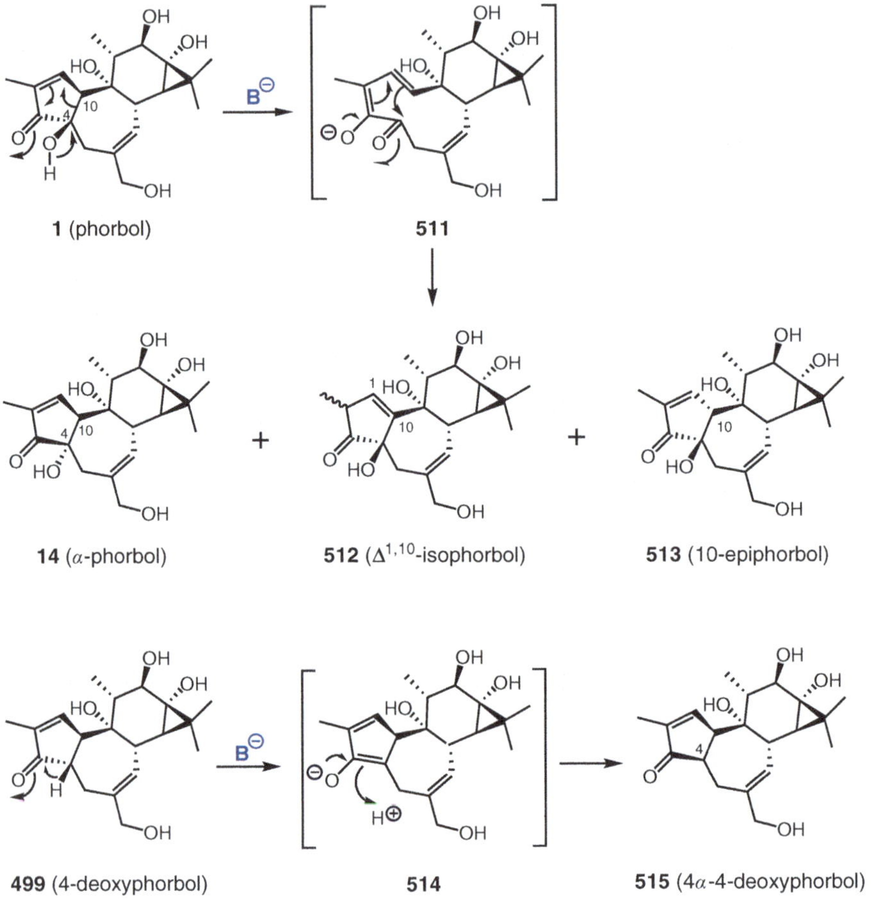

Scheme 14 Mechanism of the epimerization of phorbol (**1**) and 4-deoxyphorbol (**499**) in basic medium

or hydrogen) from a β-to an α-orientation (Scheme 1) [264]. The isomerization is irreversible, and particularly fast in compounds from the 4-deoxyphorbol (**499**) series [265]. Compounds from this series, both from the 4α- and the 4β-series, have been isolated and claimed as native, as discussed in Sect. 3.3. However, given the facility of the reaction and the complexity of their purification, they are most probably artefacts from extraction and/or isolation, and were indexed as such.

The mechanism of the epimerization at C-4 of phorbol (**1**) and 4-deoxyphorbol (**499**) derivatives is different. The reaction with phorbol is a vinylogous retro-aldol epimerization triggered by deprotonation of the 4-hydroxy group and formation of the 4,10-*seco*-dienolate **511** (Scheme 14) [265]. This next re-aldolizes with formation of a more stable *cis*-junction around the A/B rings and mainly from a conformation having H-9 oriented below the mean plane of the ten-membered ring, eventually

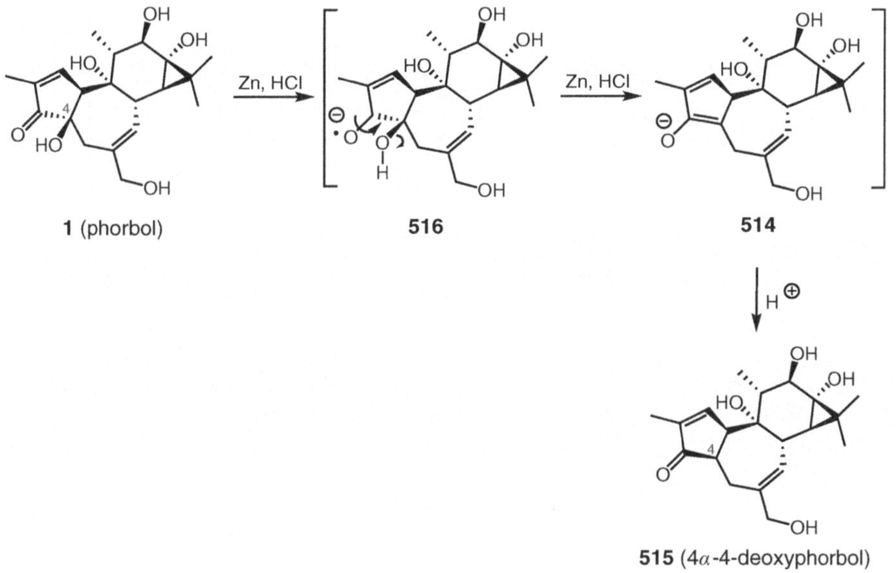

Scheme 15 Mechanism for the acyloin reduction of phorbol (**1**) to 4α-4-deoxyphorbol (**515**)

generating 4α-phorbol (**14**). The deconjugated analog from the 4β-series ($\Delta^{1,10}$-isophorbol, **512**) and the *cis*-isomer from the β-series (10-epiphorbol, **513**) were formed as minor isomerization products, originating from, respectively, an alternative protonation of the dienolate intermediate, or from a conformation of the decatriene ring with the olefin H-10 β-oriented (Scheme 14).

The epimerization of 4-deoxyphorbol (**499**) is associated with enolization at C-4 and the formation of a cyclopentadienolate intermediate (Scheme 14). The same intermediate is formed in the acyloin reduction of phorbol derivatives with zinc and acetic acid, which affords exclusively the corresponding 4α-4-deoxyphorbols (**515**) (Scheme 15) [265].

Photochemistry aside, the reactivity of 4α-phorbol (**14**) mirrors that of phorbol (**1**), with the exception of an easier acyloin deoxygenation with zinc/acetic acid to 4α-4-deoxyphorbol (**515**) [265].

Under aerobic basic conditions (NaOMe, in methanol), phorbol 20-trityl ether (**517a**) reacts with oxygen in a formal ene-reaction, affording the 2-hydroperoxy derivative **518a**. This next undergoes a Baeyer-Villiger-type rearrangement which combines the cleavage of the C-2–C-3 bond with an intramolecular attack of the 9-hydroxy on the C-3 carbonyl group. The resulting 2,3-*seco*-lactone (**519**) then undergoes isomerization of the $\Delta^{1,10}$-bond by reversible methanol addition, eventually affording the A-*seco*-derivative **520** (Scheme 16) [49]. With $Ba(OH)_2$ as a base and phorbol-20 acetate (**517b**) as a substrate, the fragmentation of the 2-hydroperoxy derivative [266, 267] is assisted by a hydroxide anion, generating a carboxylate (**521**)

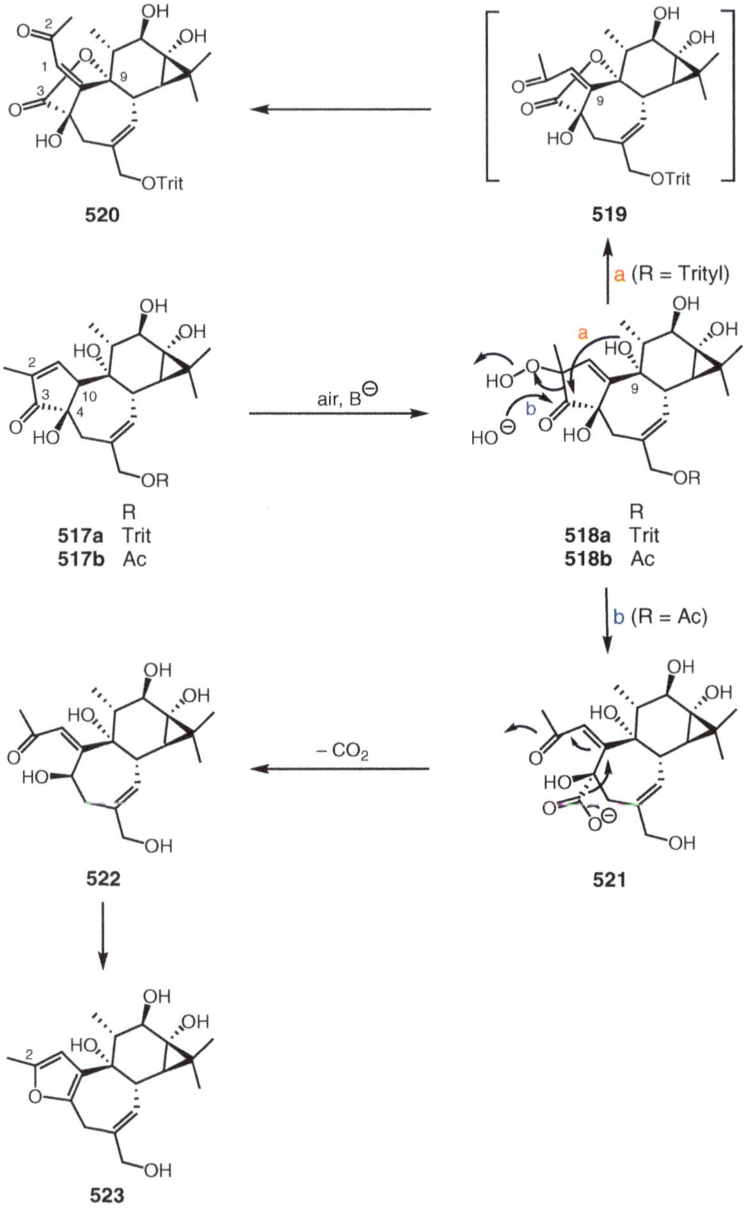

Scheme 16 Oxidative fragmentation of the ring A of phorbol 20-trityl ether (**517a**) and phorbol 20-acetate (**517b**) under basic aerobic conditions

rather than a lactone. This next loses carbon dioxide in a vinylogous fashion, eventually generating the norphorbol furan derivative (**523**) (Scheme 16) [44]. The configurational retention at C-4 is associated with the presence of a 20-trityl group, which possibly for steric reasons, slows down the retro-aldol epimerization, also giving an equilibrium mixture containing both C-4 epimers [49]. Remarkably, the 20-trityl group additionally protects the ring B Δ^6-double bond by oxygen attack, and phorbol 20-trityl ether is indeed much more stable than phorbol (**1**) and the present authors have stored this derivative for several years in a well-closed flask at a refrigerator temperature of 4°C.

Epimerization at C-4 as well as oxidative opening of ring A limit the hydrolysis/transesterification of phorbol esters under basic conditions. Apart from a careful control of pH and temperature, the reaction would therefore benefit from being carried out under an inert gas atmosphere to minimize degradation. The formation of the furan derivative **523** during the hydrolysis of croton oil is easily shown by the strong purple color this compound exhibits on TLC plates as revealed with sulfuric acid, which contrast vividly with the brown-orange color of the phorbol (**1**) and 4*α*-phorbol (**14**) zones [268].

The tertiary C-4 hydroxy group of phorbol (**1**) can be acetylated only under forced conditions (i.e. heating of phorbol 12,13,20-triacetate with acetic anhydride in the presence of *p*-toluenesulfonic acid) under which also the tertiary C-9 hydroxy group is acetylated [261]. The methylation of the 4-hydroxy group of phorbol myristate acetate (**2**) is an important reaction, since it has a detrimental effect on the interaction with protein kinase C, and the 4-*O*-methyl derivative of PMA **524** is used as negative control in biological experiments when the involvement of protein kinase C is suspected [269]. The methylation reaction is carried out in DMF, when treating phorbol esters with excess methyl iodide in the presence of Ag_2O. The reaction cannot be carried out on phorbol (**1**) itself, due to the oxidative lability of the 13-hydroxy group (see Sect. 4.3) [269]. Remarkably, under these conditions, vinylogous retro-aldol epimerization at C-4 does not take place.

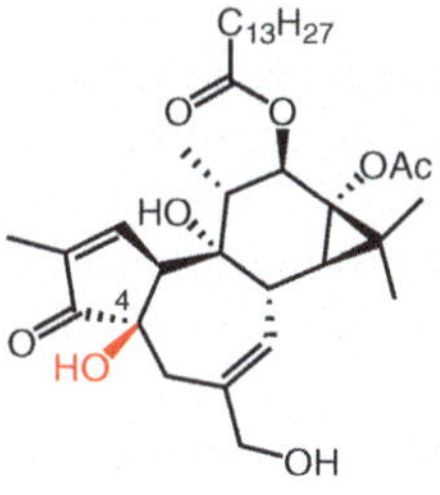

2

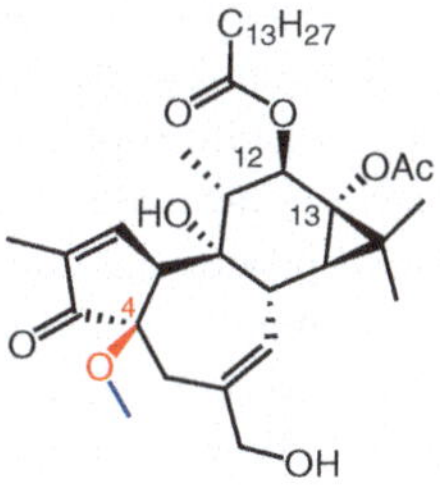

524

4.2 *The Ring B Allyl Alcohol Hotspot of Reactivity*

The reactivity associated with the ring B $\Delta^{6(7)}$ double bond also affects its three allylic carbons (C-5, C-7, and C-20). The hydroxylation of C-5 induces a reactivity profile which, just like that of the C-9 hydroxy group, can bridge the tigliane molecular regions of reactivity, and will be treated separately.

The $\Delta^{6(7)}$-Double Bond

The ring B endocyclic $\Delta^{6(7)}$-double bond is a key element of the tigliane structural motif. It has never been found in a reduced form in any naturally occurring compounds, and its main natural oxidative modification (epoxide) is rare, and always associated with β-hydroxylation at C-5, as in the tigilanoids (5β-hydroxy-6α-epoxyphorbol derivatives). Remarkably, the 5β-hydroxy-6α-epoxy functionality is instead standard in daphnane phorboids from thymelaeaceous plants [270]. Aerobic oxidation, a major oxidative degradation pathway for phorbol esters, is associated with the $\Delta^{6(7)}$-double bond and has been investigated extensively in connection with studies on the stability of phorbol myristate acetate (**2**) [271]. It was discovered that an ene reaction with single oxygen occurred spontaneously when **2** was stored in the presence of air and light, with oxygen activation being presumably triggered by traces of impurities. The mixture of degraded products included, in addition to the hydroperoxide **525**, its further transformation products **526** and **527**, as well as the *seco*-tigliane **528**, derived from the dioxetane **529** via the dicarbonyl intermediate **530** (Scheme 17). The isomeric hydroperoxide of the 6-hydroxy-$\Delta^{7,8}$-series was not detected in these studies, but compounds of this type are known to occur in their reduced hydroxy version in thymelaeaceous plants [270]. Alternative aerobic degradation pathways of phorbol myristate acetate (**2**) afforded the β-epoxide **531** and the unsaturated aldehyde **532** [271].

Tigliane esters oxygenated at C-7 and with a Δ^5-double bonds, analogs to **525**–**527**, occasionally have been reported from natural sources, but, given the presence in crude plant extracts of chlorophyll, an oxygen sensitizer, they are most likely extraction and/or isolation artefacts. Reaction with singlet oxygen and epoxidation are substrate-controlled processes, and proceed from the more accessible β-face of the phorbol myristate acetate B ring, notwithstanding the potential directing effect on the α-face that the 9-hydroxy group could exert by engaging activated oxygen with dipolar or hydrogen bonding interactions. The composition of the auto-oxidation mixture of phorbol myristate acetate (**2**) was strongly dependent on its physical state (solid/oil or solution), the nature of the solvent, and the matrix of adsorption. In solvent media, and especially acetone, the β-epoxide **531** was formed as the major product, along with the aldehyde **532** [271]. Conversely, formation of products derived from an ene-reaction (**525**–**527**) prevailed in the solid state, with the semiacetal **533** becoming the major product when **2** was adsorbed on silica gel [271]. Auto-oxidation was relatively fast at room temperature, requiring 2–3 months, but

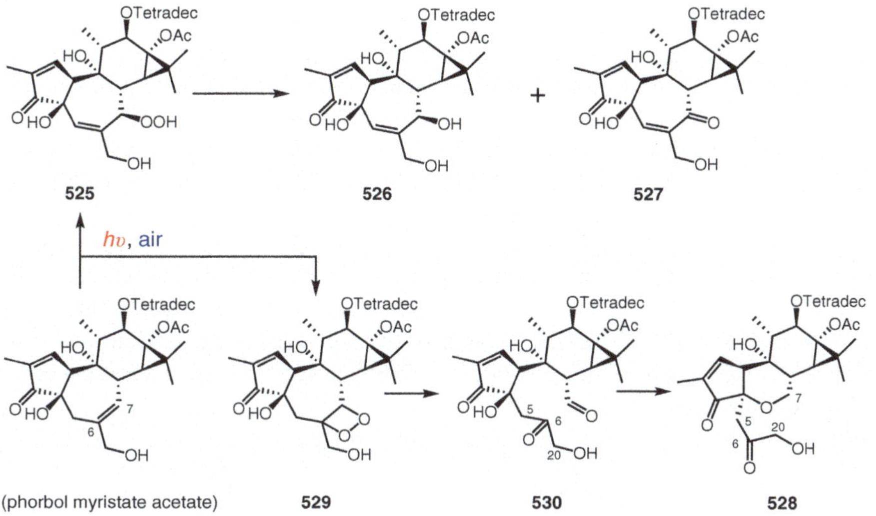

Scheme 17 Air degradation of phorbol myristate acetate (**2**) in the presence of light

was slowed significantly by cooling to −20°C. Storage in well-stoppered amber-colored vials is therefore important, possibly also under an inert gas blanket, and it has been estimated that stock solutions of phorbol esters can be stored for at least three months if maintained refrigerated (0°C or lower temperatures) and in the dark [271].

Epoxidation of the 6,7-double bond is also a major degradation pathway of phorbol esters under γ-ray radiolytic conditions [272]. The reaction was investigated in the context of studies aimed at the detoxification of waste products associated with the production of biodiesel from phorbol-ester rich oleaginous euphorbiaceous plants like *Jatropha curcas*, but the nature of the final degradation products could not be clarified [272].

For preparative purposes, the hydroperoxide **525** can be obtained by dye-sensitized photooxygenation of phorbol myristate acetate (**2**), and the epoxide **531** by epoxidation with peracids [271]. Remarkably, peracid epoxidation affords unnatural phorboid *β*-epoxides. The orientation of the oxirane oxygen is important for bioactivity, since pyramidalization of the olefin C-6 carbon affects the orientation of the primary 20-hydroxy, a group critically involved, in its dual role of hydrogen bonding-donor and acceptor, in the interaction of phorbol esters with protein kinase C [273]. In accordance with the stereochemical course of the ene-reaction, treatment of phorbol (**1**) [274], phorbol 12,13-diesters [275], and phorbol 12,13,20-triesters [275] with peracids afforded exclusively the corresponding unnatural *β*-epoxides. The reaction was first reported by Kauffmann in 1966 in the course of studies on the structure of **1** [274], and the reaction product was named phorboloxide (Phorboloxyd in German), shortened into TPA-oxide (**531**).

There has been contrasting information in the literature on the effect of epoxidation on the activity of phorbol esters. Thus, **531** was reported to be more potent than **2** in promoting radiation-induced transformation in C3H10T1/2 cells, and contamination with **531** was even suggested as a reason for the differences in potency observed between different batches of **2** [276]. However, this is in contrast with the general observation of the decreased activity of a solution of **2** on storage, and cellular and biochemically experiments of more direct mechanistic interpretation gave opposite results. Thus, complete loss of cancer-related activity was observed when phorbol 12,13-dibenzoate (**533**) was converted to its β-epoxide [275], and similar results were observed with the synthetic daphnanes *des*-epoxy-yuanhuapin (**534**) and its β-epoxide (**535**) [277]. The lack of cytotoxicity and protein kinase C affinity of the β-epoxide **535** stands in sharp contrast with the potency of the natural α-epoxide yuanhuapin (**536**) [277].

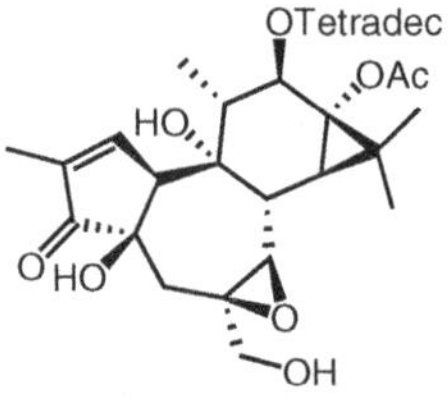

531

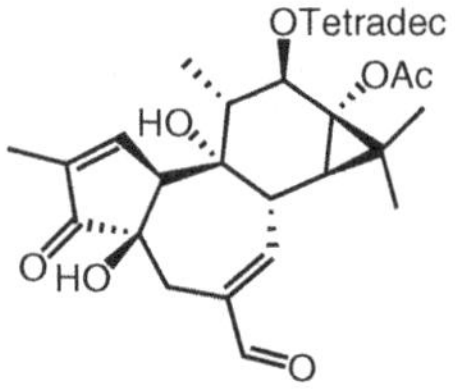

532

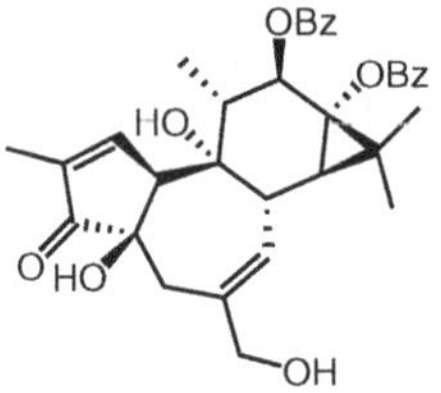

533

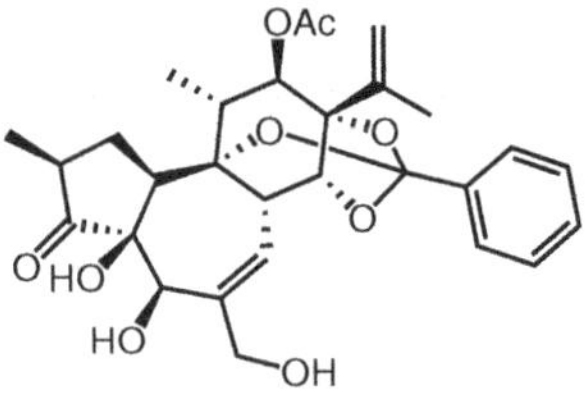

534

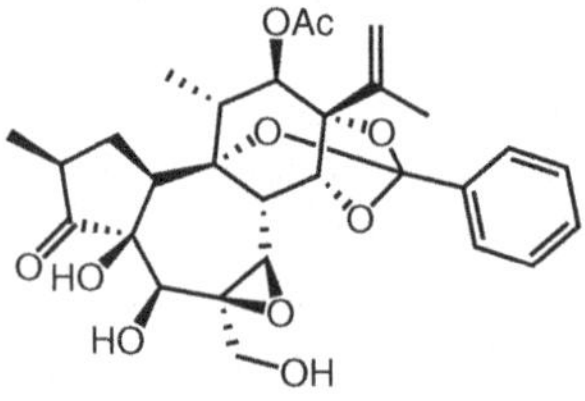

535

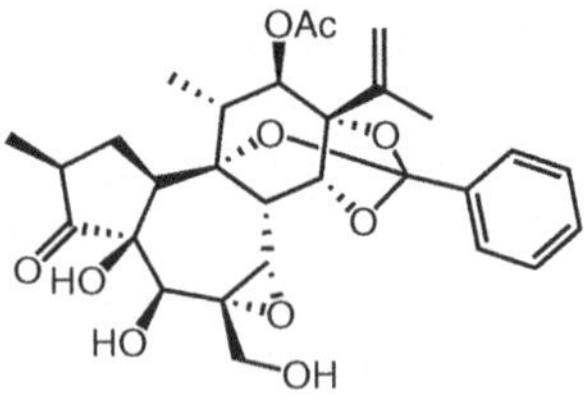

536

Since the α-orientation of the epoxide oxygen is critical for bioactivity of phorboids [275–277], a detailed study to reverse the intrinsic bias for generating unnatural and inactive β-epoxides was carried out by Wender, using phorbol 12,13-dibutyrate

(**503**) as a substrate (Scheme 18) [278]. Treatment with *m*CPBA afforded the expected β-epoxide **537** (Scheme 18). To overcome the steric bias, an alternative two-step formation of the α-epoxide **538** via the bromohydrin **540** was first attempted. Owing to the steric bulk of bromine, formation of the bromonium ion **539** is expected to take place from the convex β-face, while, for stereoelectronic reasons, attack of water should take place, at least under neutral conditions, from the α-side and on the less substituted C-7 carbon of the bromonium ion. This strategy was, however, thwarted by the propensity of bromine to oxidize the 20-hydroxy group, even when protected by esterification or by silylation, with formation of a 20-dehydro derivative **541**, de facto quenching the nucleophilicity of the adjacent double bond. A possible mechanism involves an alternative regiochemistry of electrophilic attack of Br^+, that, rather than binding to the double bond, attacks the 20-oxygen and forms a hypobromite ester, next evolving into the aldehyde **541** by β-elimination of HBr [278].

Alternative strategies based on reagent control were therefore pursued (Shi- and Sharpless asymmetric epoxidations). Epoxidation with oxone in the presence of a D-fructose-derived Shi catalyst afforded only the β-epoxide of phorbol 12,13-dibutyrate (**538**), while the L-configured enantiomer of the catalyst was unreactive, due to a mismatch between the configuration of the substrate and that of the catalyst. Better results were obtained with Sharpless epoxidation, which afforded a 1.5:1 mixture of β- and α-epoxides, validating the possibility of overcoming the β-facial selectivity under conditions of reagent control. Good results were obtained eventually with the Yamamoto epoxidation, a protocol that uses a combination of $VO(acac)_2$, *t*-butylhydroperoxide (TBHP) and a C_2-symmetric diamide as catalyst. Under optimized conditions of catalyst, solvent and temperature, a 1:9 mixture of β- and α-epoxides (**538**) eventually could be obtained is an overall 90% yield [278]. Taken together, these studies show that the double bond epoxidation of phorbol and its esters has a strong bias for the generation of unnatural and bioactivity-depleted β-epoxides. Conversely, the presence of an α-epoxide positively modulates the interaction with protein kinase C and bioactivity, but its introduction by direct epoxidation is not trivial.

An alternative multi-step α-epoxidation strategy was developed by Wender as part of the conversion of phorbol (**1**) into the anticancer agent tigilanol tiglate (**3**) by hydroxylation of the C-5 methylene and epoxidation of the adjacent double bond (Scheme 19) [45]. The double oxygenation was associated with the isomerization of the $\Delta^{6(7)}$ double bond into its Δ^5- and $\Delta^{6(20)}$ isomeric arrangements, and capitalizes on three oxidative steps and one reduction. Photo-oxygenation of phorbol 12,13-diacetate (**542**) regio-and stereoselectively afforded the Δ^5-7β-hydroperoxide **543**, and was then reduced to its corresponding alcohol **544** with thiourea [279]. Peracid epoxidation of the latter was stereoselective, and afforded the expected β-epoxide **545**, as a result of the convergence of steric (attack from the less encumbered convex β-face) and directing effects (hydrogen bonding of the 7β-hydroxy group with a peroxyacid oxygen). Tosylation of the 20-hydroxy group to generate **546** and treatment of the tosylate with NaI next triggered the transpositive deoxygenation of the glycidyl system, with the formation of the bis-allylic $\Delta^{6(20)}$ alcohol **548**. The reaction is presumably a retro-type I Ferrier reaction [280], with the fleeting 20-iodide **547**

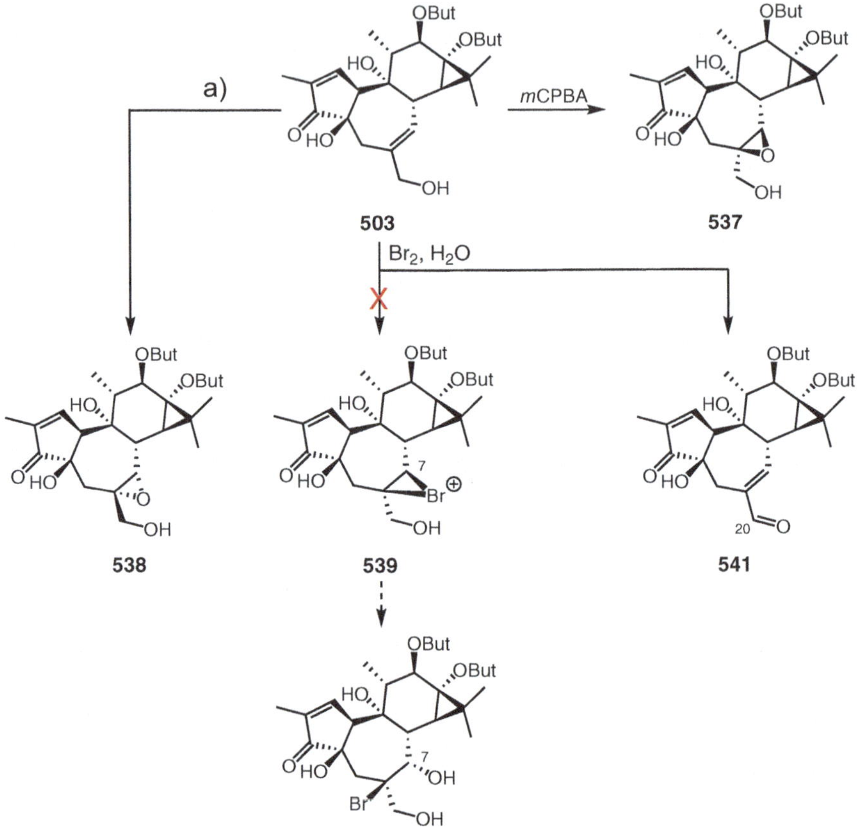

Scheme 18 Epoxidation studies on phorbol-12,13-dibutyrate (**503**). a) 10% VO(O*i*-Prop)$_3$, 20% ligand ((*R,S*)-*N,N*′-dihydroxy-*N,N*'-3,3-diphenylpyhenylacetyl-*trans*-1,2-cyclohexanediamine), 2 equiv. TBHP, C_6H_6, 4°C; yield 90% ((1:9 mixture of *β*-epoxide and *α*-epoxide (**538**))

as a substrate. The reductive opening of the epoxide was then triggered by the attack of an iodide anion on the 20-iodine atom. This generated molecular iodine and a C-5 alcoholate (Scheme 19), formally reversing the electron movements involved in the type I Ferrier reaction, a Lewis acid-mediated cationic process producing an intermediate oxonium ion [280].

The bis-allylic diol **548** was then isomerized regioselectivity with rhenium (VII) oxide [281] to the endocyclic $\Delta^{6(7)}$-*β*-hydroxy derivative **549**, with the driving force of the reaction being the higher stability of a trisubstituted endocyclic bond compared to a disubstituted exocyclic one. The peracid epoxidation of **549** gave the expected *β*-epoxide, but the formation of the six-membered C-5,20-acetonide **550** remarkably reversed the regiochemistry of the epoxidation, affording in excellent yield the *α*-epoxide **551** when *m*CPBA was replaced by dimethyldioxirane, a less encumbered

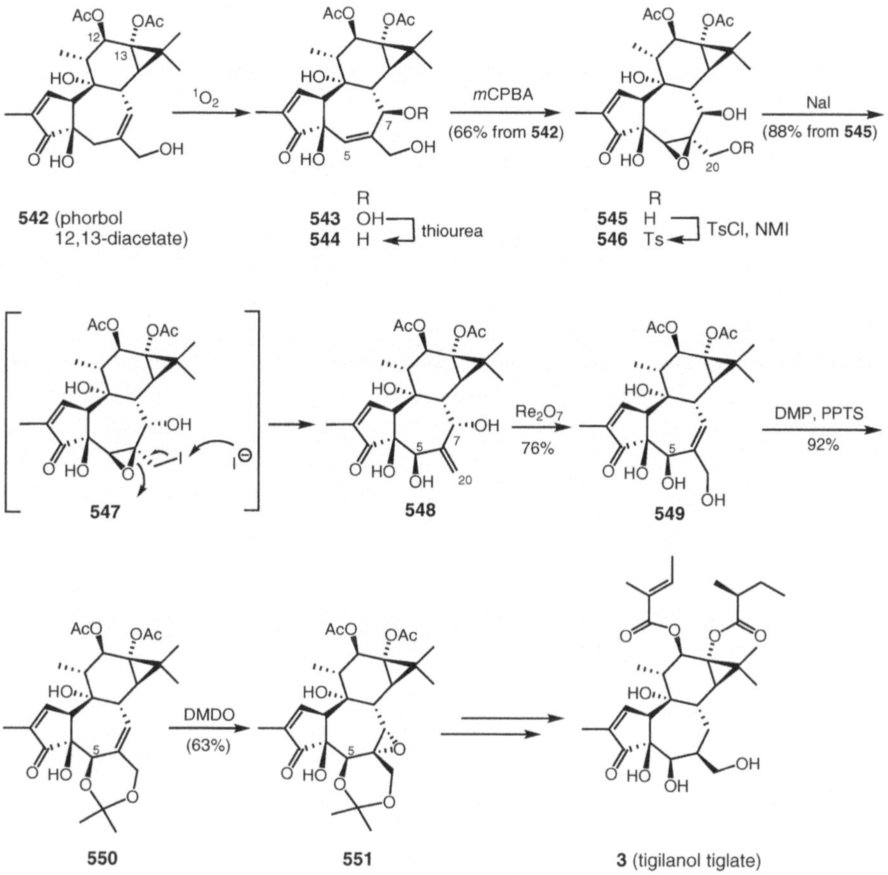

Scheme 19 Conversion of phorbol diacetate (**542**) into tigilanol tiglate (**3**). TsCl = *p*-tosyl chloride; NMI = *N*-methylimidazole; DMP = dimethoxypropane; PPTS = pyridinium *p*-toluene sulfonate; DMDO = dimethyldioxirane

and more reactive reagent. Completion of the semi-synthesis of tigilanol tiglate (**3**) from phorbol then involved the uneventful remodulation of the ring C acylating decoration and final deprotection [279].

The allylic reactivity of the $\Delta^{6(7)}$ double bond is associated with the functionalization of C-7 and its Δ^5 relocation, as shown in Scheme 18. Attempts to functionalize the allylic carbons with retention of the $\Delta^{6(7)}$ position have so far failed [44, 282]. Thus, allylic oxidation of phorbol 12,13,20-triacetate (**552**) with selenium (IV)-reagents (selenium dioxide, selenium dioxide/*t*-butylhydroperoxide, phenylseleninic anhydride) gave the 20-aldehyde **553** as the only reaction product, echoing the Wender results on bromine addition to the $\Delta^{6(7)}$-double bond discussed previously. Removal of the protecting group of the 20-hydroxy group and its oxidation were also observed when a trityl group replaced the 20-acetyl.

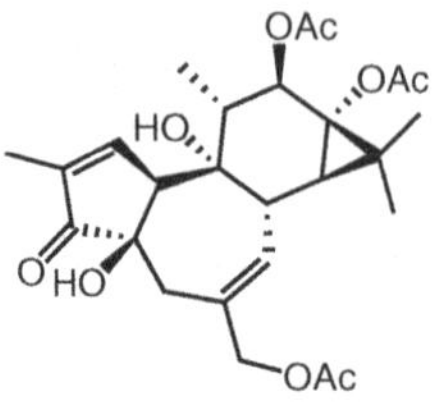

552 (phorbol 12,13,20-triacetate)

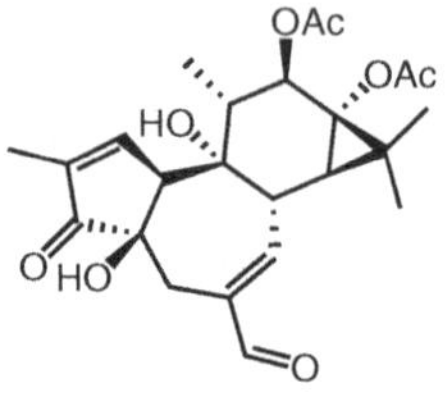

553 (phorbol 20-aldehyde)

A notable rearrangement involving the $\Delta^{6(7)}$ double bond was observed instead when the acetonide of 3αH-dihydrophorbol 12,13,20-triacetate (**554**) was treated with pyridinium chlorochromate and sodium acetate (Scheme 20). Initial formation of the cyclic chromium (IV) diester **555** was followed by displacement of the chromium ester bond at C-6 by the 9-hydroxy group, with generation of the chromium(IV) ester **556**, eventually evolving into the C-8 ketone **557** [282]. The intramolecular photochemistry of the $\Delta^{6(7)}$ double bond has been discussed in Sect. 4.1.

The $\Delta^{6(7)}$ double bond of phorbol 12,13,20-triacetate (**552**) has been dihydroxylated chemoselectively with OsO_4 in pyridine [283]. In accordance with the stereoselectivity observed in the epoxidation, attack occurred exclusively from the β-face, with formation of the 6β,7β-glycol **558**. Treatment of this compound with lead

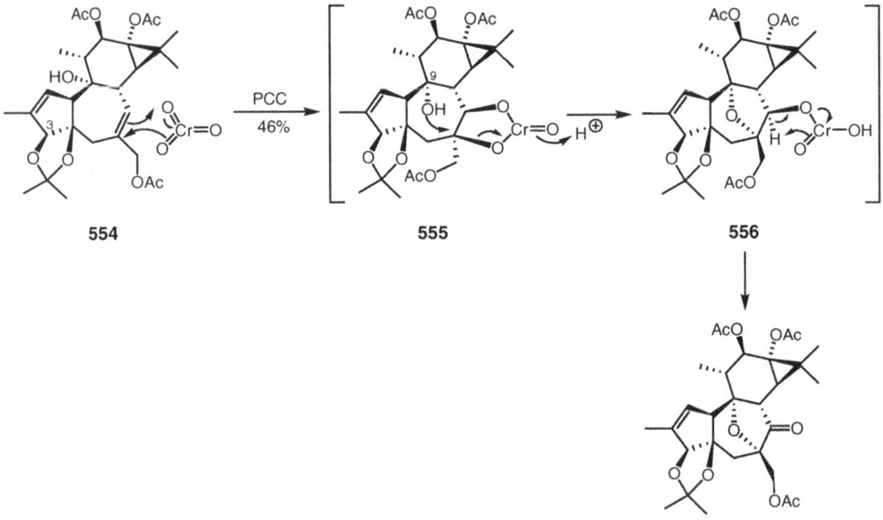

554 **555** **556** **557**

Scheme 20 Reaction of 3αH-dihydrophorbol triacetate 3,4-acetonide (**554**) with Cr(VI) reagents. In the mechanistic scheme, pyridinium chlorochromate (PCC) is simplified into chromium(VI) anhydride

tetraacetate or sodium periodate cleaved ring B, with formation of the 6,7-*seco*-tigliane **559**, for which the C-7 aldehyde carbonyl engaged the 4-hydroxy group into a δ-lactone ring. Historically, this reaction was critical for the localization of the single unfunctionalized methylene of phorbol (**1**), which evidently was adjacent to the non-conjugated double bond [283], contributing to the rejection of the formula where this methylene had been located on the five-membered ring (see Sect. 2).

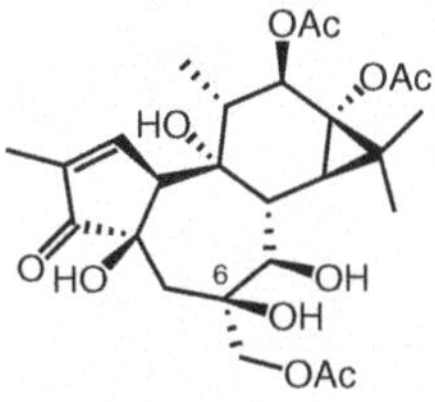

558

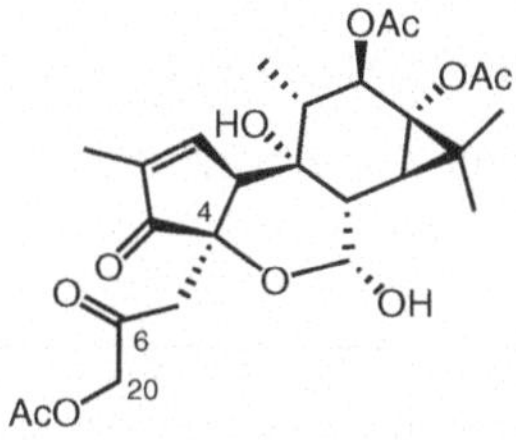

559

Due to the presence of an allylic hydroxy group, hydrogenation of the ring B double bond is generally accompanied by deoxygenation of C-20, even when working at hydrogen pressure [284]. Hydrogenation shows, however, selectivity for the $\Delta^{6(7)}$ double bond compared to the ring A enone system.

The C-20 Hydroxymethylene Group

The 20-hydroxy group is not able to be directly acetylated in a selective way, since acetyl chloride and acetic anhydride cannot discriminate the primary allylic C-20 hydroxy and the tertiary cyclopropyl C-13 hydroxy groups [261]. Due to its enol-like nature, the latter is highly reactive, and even can be acetylated chemoselectively in the presence of the 20-hydroxy group if strictly stoichiometric amounts of the acylating reagents are used and the reaction is carried out at low temperature [261].

Conversely, and in blatant defiance of the principle of microscopic reversibility, a 20-ester group can be removed selectively in the presence of 12- and 13-esters by acidic transesterification with perchloric acid in methanol, making it possible to synthesize 12,13-phorbol diesters from their corresponding 12,13,20-triesters [261]. Alternatively, the 20-hydroxy group can be protected by tritylation or silylation, the 12- and 13-hydroxy groups are esterified, and eventually the protecting group (trityl or silyl) of the 20-hydroxy group is removed [263]. A vis-à-vis comparison of the two strategies was reported for the synthesis of phorbol 12,13-diacetate (**542**) by Wender, and the three-step protocol (protection-acetylation-deprotection) gave a better yield than the two-step protocol (acetylation-selective deacetylation) (82% vs. 65%) [279].

Quaternization of C-7 as in lumiphorbol (**500**) quenches the selective reactivity of the 20-ester [263], which is, however, retained in tigilanoids, where the inductive effect of the electronegative oxirane oxygen seemingly compensates for the increased steric congestion associated with the quaternization of C-6 [8]. A surprising and disturbing loss of the 20-protecting group followed by oxidation of the 20-hydroxy

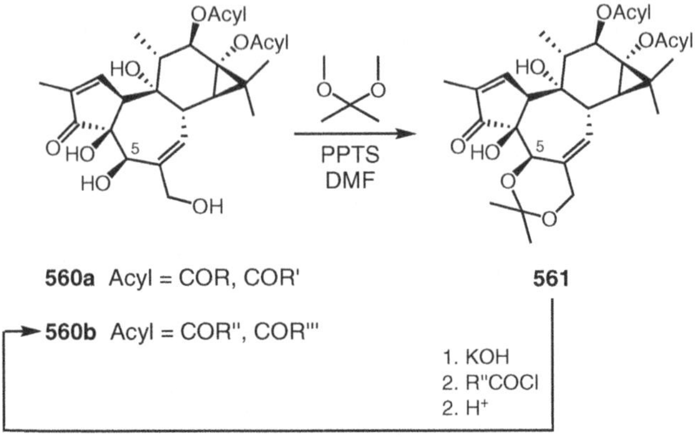

Scheme 21 Reversible protection of the 5β,20-diol system of tigilanoids as exemplified by **561**

group was observed during the addition of bromine to the 6,7-double bond and during its allylic oxidation, as discussed in the previous section [279].

Alkylation of the 20-hydroxy group is complicated by the instability of the vinylogous aldol system at the A/B ring junction under basic conditions, leading to the formation of 20-alkyl 4α-phorbol derivatives. Acetalization with a 5β-hydroxy group is, however, easy under acidic conditions with both acetone and 2,2-dimethoxypropane, and the mild acid pyridinium *p*-toluenesulfonate proved sufficient for the reaction [8, 279]. Deprotection with diluted strong acids (perchloric, *p*-toluenesulfonic) is then possible, although care should be taken not to trigger cationic rearrangements associated with the cyclopropylic C-13 ester group [8] (see also Sect. 4.4). Capitalizing on the reversibility of acetonide formation, the ring C acylation profile ca be modulated, converting diesters of the **560a** type into diesters of the **560b** type via the acetonide **561** (Scheme 21) (see Sect. 4.4).

The 20-hydroxy group could be replaced by a chlorine by reacting phorbol 12,13-diacetate (**542**) with mesyl chloride in pyridine. The resulting 20-deoxy-20-chloro derivative (**562**) next could be dehalogenated reductively with Raney nickel, which occurs more easily compared to the deoxygenation of the 20-acetate [32]. Surprisingly, the corresponding 12,13-didecanoate was unreactive under these conditions [32].

An unexpected reaction occurred when tigilanol tiglate (**3**) was subjected to the Appel reaction with CCl_4 and triphenylphosphine (TPP) [8]. Rather than the expected 20-deoxy-20-chloro derivative, its 4,5-acetal with triphenylphosphine oxide **15** was obtained in 19% yield. The X-ray analysis of **15** was useful in confirming the structure of tigilanol tiglate (**3**) and in securing the configuration of its α-methylbutyric ester group. The unusual course of the Appel reaction can be rationalized in terms of intramolecular attack of the 5-hydroxy group on the oxyphosphonium intermediate **563**, produced from the interaction of the 20-hydroxy with the phosphonium salt generated by the interaction of triphenylphosphine and CCl_4. This could originate from the interaction of the 20-hydroxy with the phosphonium salt generated from

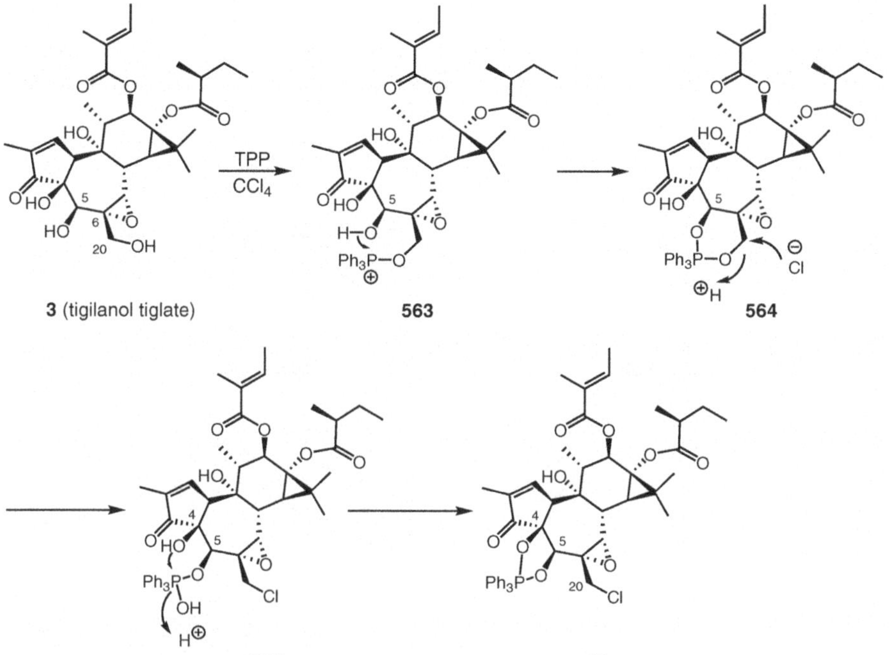

Scheme 22 Unusual course of the Appel reaction on tigilanol tiglate (**3**)

triphenylphosphine and CCl_4. Next, the phosphine oxide acetal **564** underwent attack by a chloride at the C-20 primary carbon, with the departure of a phosphine oxide semiacetal, eventually acetalized by the C-4 hydroxy group via **565** to yield **15** (Scheme 22).

Due to its allylic nature, the 20-hydroxy group of phorbol 12,13-diesters can be oxidized cleanly with activated MnO_2, and a high-yield reaction was also observed with Cr(VI)-based reagents [285]. The reaction also takes place on aerial exposure of phorbol esters having a free 20-hydroxy group, and 20-aldehydes are some of the major auto-oxidation products of phorbol 12,13-diesters [271]. The 20-aldehyde can be reduced chemoselectively with labeled (deuterium, tritium) borohydrides without affecting the less reactive enone carbonyl [285]. The reaction provides access to isotopically modified analogues important for cellular and pharmacokinetic studies, due to their selective detectability. Thus, 20($^2H_3$)-phorbol dibutyrate (**566**) has played a major role in the identification of protein kinase C as the molecular target of phorbol esters [285]. Of historical relevance is also the reaction of 20-dehydrophorbol-12,13-diacetate (**567**) with 2,4-dinitro-5-iodophenylhydrazine, since the resulting highly crystalline hydrazone (**568**) could be solved by X-ray crystallography, confirming also in terms of absolute configuration the structure established on a neophorbol derivative (see Sect. 2) [286]. By dehydration of the 20-oxime **569**, the cyano derivative **570** could be obtained [16], while oxidation of the 20-formyl group with selenium dioxide and hydrogen peroxide afforded the corresponding 20-carboxylate **571** [287].

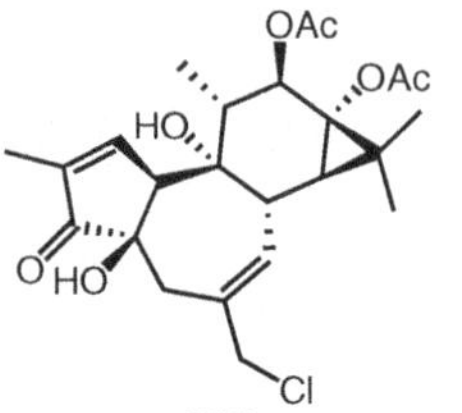

562

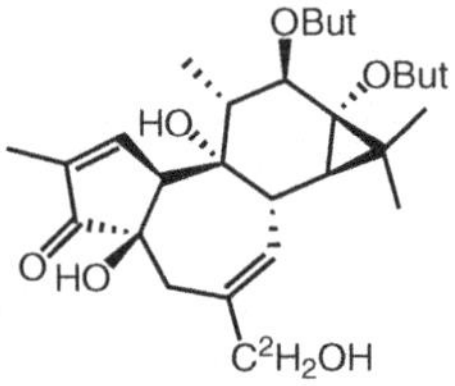

566

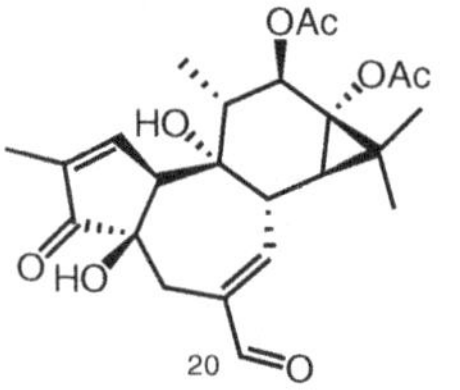

567

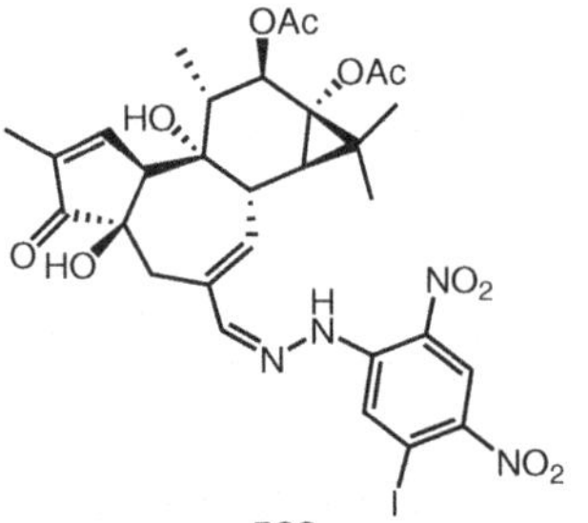

568

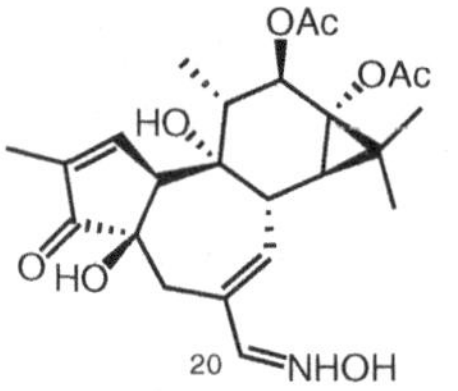

569

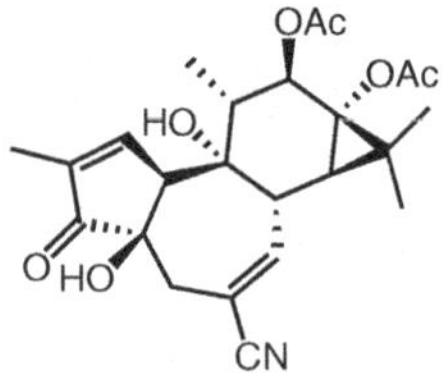

570

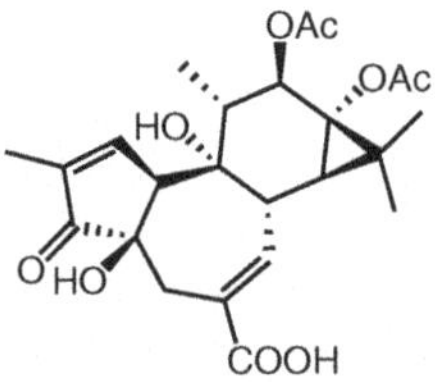

571

The 20-hydroxy group could be removed reductively by treatment of phorbol 12,13,20-triacetate (**552**) with Raney nickel in ethanol or methanol [287]. The reaction has been reported to require careful control to avoid reduction of the adjacent double bond, but the experimental conditions have not been described in detail [287]. Hydrogenolysis of the 20-hydroxy group was observed in the hydrogenation of phorbol 12,13,20-triacetate (**552**) with palladium on charcoal or platinum as catalyst [26]. The uptake of three moles of hydrogen could be eliminated using Adams catalyst (platinum(IV) oxide hydrate), although the product obtained was not a derivative of tetrahydrophorbol, but of 6,7-dihydro-20-deoxyphorbol (**572**) [30]. All these compounds presumably are inactive in terms of protein kinase C binding, since hydrogenation of croton oil has been reported to eliminate its sharp taste as well as its toxicity [288]. Elimination of the C-20 oxygenated function of phorbol triacetate under basic conditions occurs in a vinylogous fashion, affording the $\Delta^{6(20),7}$-diene **573** [32].

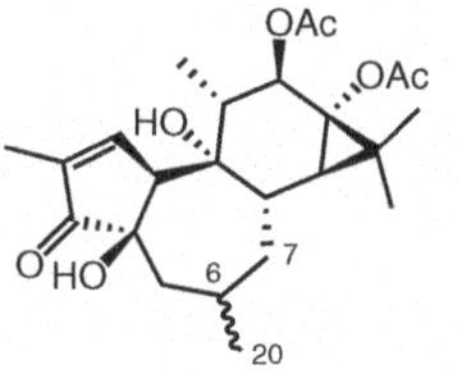

572

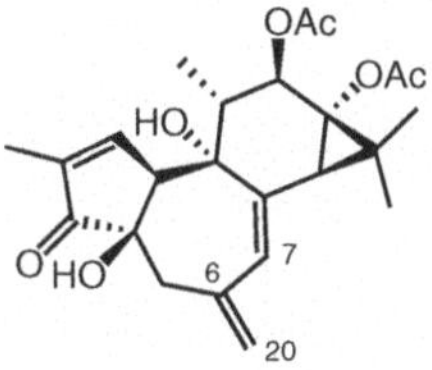

573

4.3 The Glycol Group on the C/D-Ring System

The C-12 Hydroxymethine Group

In acylation reactions, the C-12 hydroxy group shows the reactivity expected for a secondary alcohol [261]. It is therefore less reactive than the primary 20-hydroxy for steric reasons, but also less reactive than the tertiary 13-hydroxy group, and for the latter its reactivity is magnified by a cyclopropylic nature (see Sect. 4.3). Given these differences, the acylation of phorbol (**1**) may occur in a step-wise fashion, making it possible to prepare polyesters with a tailored acylation profile. Interestingly, the esterification profile of natural phorbol polyesters is generally diversified, as discussed in Sect. 3.3. In a transesterification reaction, acyl rearrangement with migration of ester groups from the secondary 12-hydroxy to the tertiary, but more reactive, 13-hydroxy group does not occur at a significant rate [261], and the same also holds for the reverse reaction, namely, the migration of the tertiary 13-esters to the secondary 12-hydroxy group [261]. This lack of acyl rearrangement greatly simplifies the planning of a specific acylation profile by avoiding scrambling of ester groups around the glycol system.

Under acidic conditions, the protonated 12-hydroxy forms an oxonium ion that dehydrates easily, generating a stable cyclopropylcarbinyl cation which eventually evolves into a mixture of rearranged products [32, 289] The reaction, known as the Flaschenträger reaction [19], dates from 1934, and was conducted originally by treating phorbol (**1**) with hot diluted sulfuric acid [19], while more modern studies were done using hydrochloric acid in dry ether [30, 289]. The C-12 cation generated in the Flaschenträger reaction (**574**) can undergo a Tiffeneau-Demjanov-type [290]

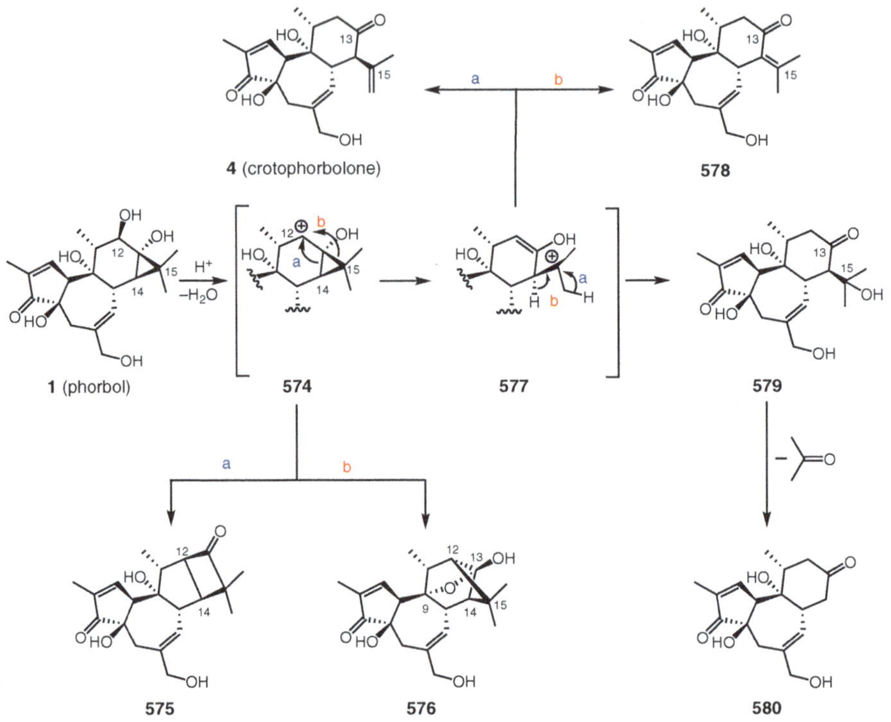

Scheme 23 Mechanistic view of the Flaschenträger reaction of phorbol (**1**)

ring enlargement (semipinacol-type rearrangement), with migration of either of the two cyclopropane bonds centered on C-13 (Scheme 23). Formation of a protonated cyclobutanone carbonyl of the 14(13 → 12)-*abeo* prevails over formation of its isomer from the 15(13 → 12)-*abeo* series, with eventual evolution into, respectively, phorbobutanone (crotophorbolone K (**575**)) and isophorbobutanone (**576**), and the latter existing as a cyclic 9,13-hemiacetal. Alternatively, the 12-cyclopropylcarbinyl cation **574** can undergo homoallyl rearrangement to the tertiary C-15 cation **577**, and is then quenched by proton loss from one of the geminal C-15 methyls and tautomerization to the rhamnopholanes crotophorbolone (**4**) and isocrotophorbolone (**578**), or, alternatively, by water trapping to their hydrate derivative hydrocrotophorbolone (**579**). Under the acidic conditions of the reaction, this undergoes retro-aldol fragmentation to the tris-*nor*-tigliane **580** and acetone.

Due to its highly crystalline state, crotophorbolone (**4**) was the first compound characterized from the Flaschenträger reaction [291], and has played an important role in the structure elucidation of phorbol (**1**) (see Sect. 2). For preparative purposes, **4** can be prepared from **1**, or its 13,20- and 12,13,20-triesters. Treatment of **1** under the classic conditions of the Flaschenträger reaction (90 min refluxing in 0.02 *N* H_2SO_4 methanol-water, 7:2) leads to a mixture of crotophorbolone (**4**) and phorbobutanone (= crotophorbolone K (**575**)). The former can be crystallized directly from the crude mixture of products obtained after quenching the reaction with $NaHCO_3$ and evaporation, while purification of the latter requires chromatography or a careful series of fractionated crystallizations [30, 289]. The overall yield was reported in the range of 30% for crotophorbolone (**4**) and 15% for phorbobutanone (**575**), and compounds **576**, **578**, and **579** were characterized as minor isomerization products [289]. Treatment of phorbol 12,13,20-triacetate (**552**) with the complex BF_3-acetic acid in ether afforded the addition product of acetic acid to 20-acetylcrotophorbolone (acetoxycrotophorbolone 20-acetate (**581**)) in modest yield (ca. 25%) [30], while refluxing in ether with $AlCl_3$-$LiAlH_4$ gave, after acetylation of the rection mixture, the non-conjugated enolacetate of crotophorbolone-20-acetate (**582**) in high yield (ca. 75%) and as a single reaction product, which can be converted to crotophorbolone (**4**) by methanolysis [30]. The enone carbonyl was not affected in the reaction, possibly because of the formation of an insoluble complex in the course of the reaction, while the formation of the non-conjugated enolacetate is unclear mechanistically. The $AlCl_3$-$LiAlH_4$ combination is the only acid that gave a single major compound using the Flaschenträger reaction [30].

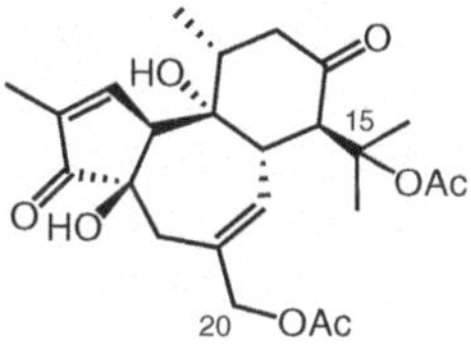

581

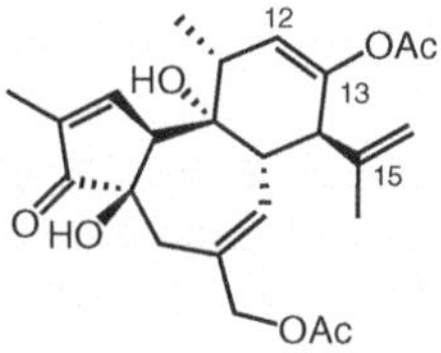

582

Formation of phorbobutanone (**575**) can be avoided if the Flaschenträger reaction is carried out under neutral conditions on phorbol 13,20-diacetate (**583**), but acetoxycrotophorbolone-20 acetate (**582**) is produced as by-product [291]. Phorbol 13,20-diacetate (**583**) can be prepared directly from phorbol, capitalizing on the higher reactivity of the 13- and 20-hydroxy groups compared to that at C-13 (see Sect. 4.2) [9]. Treatment with 2 equivalents of mesyl chloride in pyridine at 0°C and methanolysis of the excess mesyl chloride, leads via the 12-mesylate **584**, to the formation of a mixture of 54% enol acetate of crotophorbolone 20-acetate (**582**) and 20% acetoxycrotophorbolone 20-acetate (**581**), the result of internal quenching of the C-15 cation by the enol acetate. After separation, Zemplen methanolysis of **582** afforded the parent diterpenoid crotophorbolone (**4**) [30].

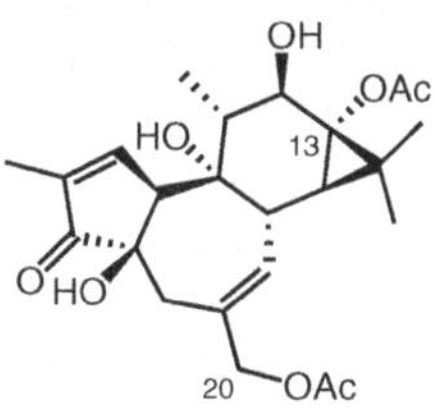

583

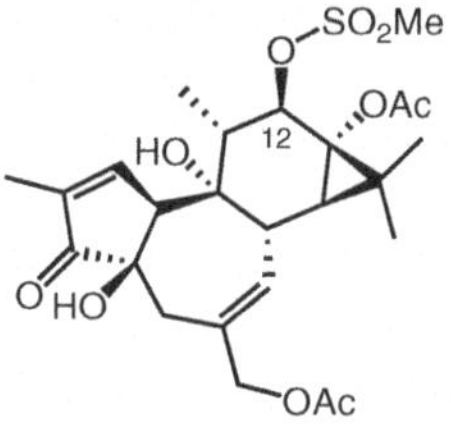

584

Formation of the mixture of rearranged products associated with the Flaschenträger reaction was also observed when phorbol was incubated anerobically with human intestinal bacteria [292]. Given the easy formation of the C-12 tiglianyl cation, it can not be ruled out that acidification and not enzymatic activity was responsible for the formation of the rearranged products.

Crotophorbolone (**4**) has been isolated as a trace constituent of *Euphorbia fischeriana* Steud. [183], but might well be an artefact of the long fractionation of the plant extract that led to its isolation, with its biological profile being unimpressive. Nevertheless, it has become a target for total synthesis, two of which two have been reported [293, 294]. From a less academic perspective, **4** was used by Wender for a semi-synthesis of the anti-HIV agent prostratin (13-acetyl-12-deoxyphorbol (**192**)) [295]. Although only the removal of the C-12 hydroxy group is involved in the conversion of phorbol-13 acetate (**585**) to prostratin (**192**), the reaction can not be carried out directly, due to the tendency of the cyclopropyl group to open under acidic-(Flaschenträger reaction) and radical conditions (see below). An indirect strategy, involving opening and then closure of the cyclopropane ring was developed (Scheme 24). For this purpose, crotophorbolone (**4**) was reacted with hydrazine, and the resulting hydrazone **586** was converted into the corresponding pyrazoline (**587**) by heating in pyridine-diisopropylethylamine at 150°C in a formal Markovnikov addition of the distal amino group of the hydrazone to the exomethylene. The chemoselectivity of the reaction with hydrazine shows that the C-3 carbonyl of crotophorbolone (**4**) shares the same unreactivity of the corresponding carbon of phorbol (**1**) (see Sect. 4.1). Lead(IV) tetraacetate oxidation of the pyrazoline afforded the acetoxy diazene derivative **588** and oxidatively introduced the acetoxy group at C-13, with photolytic nitrogen extrusion completing the conversion of phorbol (**1**) to prostratin (**192**). This synthesis scheme could be diversified to produce analogues of prostratin modified at the C-13 oxygen function. Thus, the use of different lead(IV) salts afforded the corresponding 13-esters, while a change to a iodine(III) oxidant ($PhIAc_2$) step resulted in the solvolytic introduction of an ether group from the solvent used in the oxidation reaction. Unlike the introduction of the acetyl, there was no stereocontrol in the introduction of alkoxy groups, but the photolytic step afforded convergently the same cyclopropane derivative from both C-13 isomers [295]. This chemistry, developed by Wender was then used in the two total syntheses of prostratin reported to date (see Sect. 5).

The 12-hydroxy group of phorbol 20-trityl ether (**517**) could be etherified by treatment with diazoalkanes (diazomethane or diazoethane) in the presence of aluminum isopropylate [269]. Williamson-type alkylations are not suitable for phorbol, given its instability to basic conditions (see Sect. 4.1). The reductive removal of the 12-hydroxy group has so far failed, since the classic deoxygenation reactions involve the formation of an alkylcyclopropyl radical, a radical "clock" that rearranges to an allyl radical by cyclopropane ring opening at near diffusion-controlled rate, as exemplified by the reaction of the 12-thioimidazolide of 20-trityl phorbol-20-acetate (**589**) [295]. When this compound was subjected to the Barton-McCombie deoxygenation (heating with $(n\text{-Bu})_3SnH$ in toluene in the presence of azoisobutyronitrile (AIBN)),

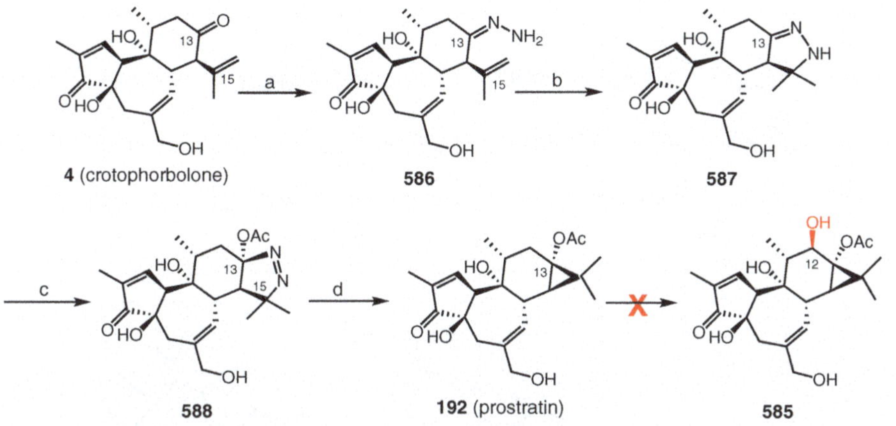

Scheme 24 Conversion of crotophorbolone (**4**) into prostratin (**192**). (a) $N_2H_2.H_2O$, AcOH. (b) pyridine, diisopropylethylamine, Δ. (c) $Pb(OAc)_4$ (overall 49% from **4**). (d) *hν* (67–92%)

a mixture of rhamnopholanes from the 4*α*- and 4*β*-series was obtained (**592**) via the radicals **590** and **591** [295] (Scheme 25).

Oxidation of the 12-hydroxy group of phorbol 13,20-diacetate (**583**) afforded the corresponding 12-ketone (**593**), for which the hydride reduction gives mainly compounds from the 12-*epi*-phorbol series (**594**) as a result of hydride attack from the less hindered *β*-face of the carbonyl, *trans* to the C-11 methyl [296, 297], as will be detailed in the discussion of Wender's total synthesis of phorbol (see Sect. 5.1). The 12-carbonyl is less reactive than the enone C-3 carbonyl on hydride reduction. Thus, hydride reduction of 12-dehydrophorbol 13,20-diacetate (**593**) chemoselectively afforded neophorbol 13,20-diacetate (**595**) [298], the isomer of phorbol 13,20-diacetate (**583**), having a reversed functionalization at C-3 (hydroxy vs. carbonyl) and C-12 (carbonyl vs. hydroxy) [298]. X-Ray data analysis on the 3-*p*-bromobenzoyl derivative of neophorbol 13,20-diacetate (**8**) was critical in establishing the structure of the natural product, more than forty years after its isolation [29]. The decreased

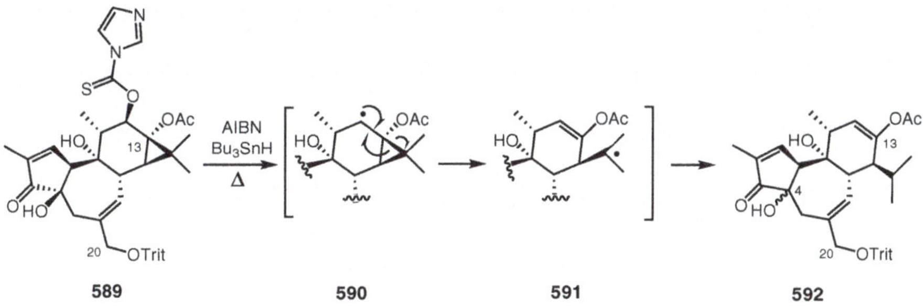

Scheme 25 Attempted C-12 deoxygenation of the thioimidazolide **589** by radical chemistry

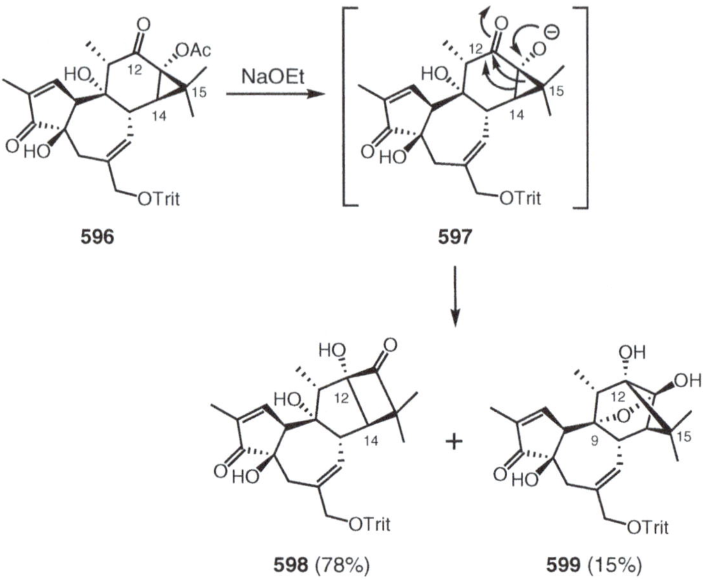

Scheme 26 α-Ketol rearrangement of the 12-dehydrophorbol derivative **596**

reactivity of the C-12 carbonyl could be due to carbonyl stabilization from the cyclopropane ring. The bent cyclopropane C–C bonds are excellent hyperconjugative donors due to their increased *p*-character, and can stabilize an adjacent carbonyl by $\sigma_{C\text{-}C} \rightarrow \pi^*_{C=O}$ donation [299]. This stabilization is lost in the tetrahedral intermediate resulting from nucleophilic addition to the carbonyl, of which the reactivity is therefore reduced.

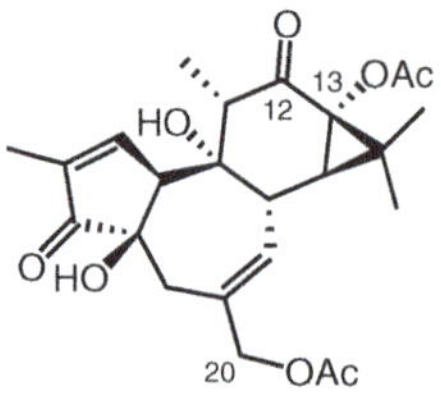

593

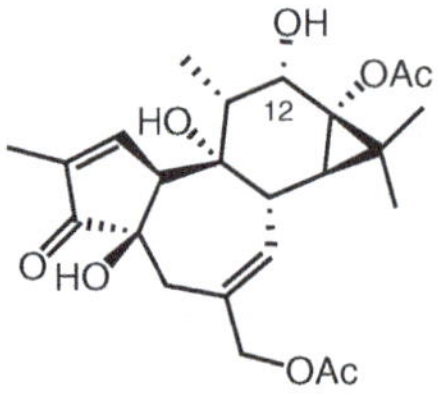

594

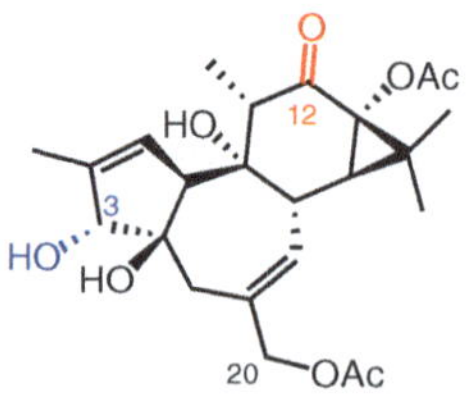

595

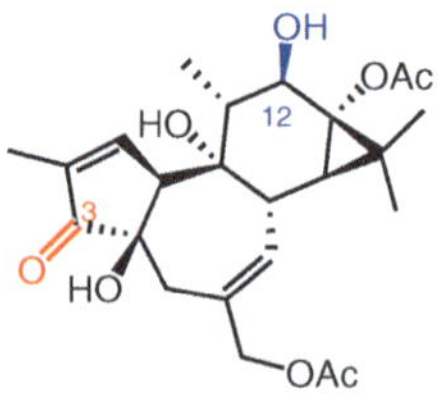

583

Under slightly basic conditions, 20-trityl-12-dehydrophorbol 13,20-diacetate (**596**) underwent an α-ketol rearrangement via **597**, giving a mixture of the corresponding compounds of the 12-hydroxyphorbobutanone (**598**) and 12-hydroxyisophorbobutanone (**599**) series (Scheme 26) by migration to C-12 of, respectively, C-14 and C-15, in a process similar to that observed in the Flaschenträger reaction for the formation of the corresponding 12-deoxygenated compounds (phorbobutanone (**575**) and isophorbobutanone (**576**) shown in Scheme 23) [298]. Under the slightly basic conditions of the reaction, no *trans-cis* isomerization of the A/B ring junction to compounds of the 4α-series was observed [298].

The Tertiary Hydroxylated C-13 Cyclopropyl Carbon

Due to its cyclopropyl nature, the tertiary C-13 hydroxy group shows unusual reactivity in oxidation and acylation reactions. Phorbol derivatives with a free 13-hydroxy group give positive tests with the Tollens and the Fehling reagents [300], a reaction typical of secondary α-ketols and at that time unprecedented for tertiary α-ketols, but now recognized as general for cyclopropanols, a class of compounds prone to oxidative ring opening [246, 301]. Etherification of the 12-hydroxy group does not affect the positivity of the test, ruling out the involvement of the 12-hydroxy in the reaction [296]. The oxidized compounds formed in the test have never been isolated and characterized, and phorbol (**1**) is also oxidized by silver oxide and lead dioxide. Under the conditions of the Tollens and Fehling tests, a C-13 oxygen radical could be formed, next evolving into oxidatively cleaved products not unlike those obtained with other metal oxidants (lead(IV) tetraacetate, sodium bismuthate, see below).

The reaction of phorbol (**1**) with lead tetraacetate is of great historical relevance in the context of the structure elucidation of phorbol (see Sect. 2). The reaction was first reported by Kauffmann, who described the formation of acetone and a cleaved compound named tigl(i)ophorbol (**604**) [24]. The reaction was carefully investigated by Crombie [30] and by Hecker [302]. The formation of acetone could not be confirmed, and, in addition to tigl(i)ophorbol (**604**), other compounds were obtained from the reaction, all retaining the twenty-carbon skeleton of the natural product. These compounds were named tigl(i)ophorbol A (bisdehydrophorbol (**604**)), tigl(i)ophorbol B (**605**) and tigl(i)ophorbol T (phorbolactone semiacetal (**603**)). The dual semantic is the result of a different assignment by the two groups who investigated the reaction. The composition of reaction products depended on the solvent and the amount of lead tetraacetate used, with **604** and **605** forming in similar amounts with stoichiometric amounts of oxidant. With an excess of oxidant, tigl(i)ophorbol T (phorbolactone semiacetal (**603**)) became the major reaction product. The same mixture of compounds was obtained with sodium periodate and sodium bismuthate [303], implying a similar mechanism. Rather than the expected cleavage of the C-12–C-13 bond, these oxidations occurred with cleavage of the C-13–C-15 cyclopropane bond (Scheme 27). The formation of tigl(i)ophorbol (**604**) could be the result of a process initiated by the involvement of the more acidic 13-hydroxy into a lead(IV) alcoholate (**600**, bismuth alcoholate or iodine ester for the other

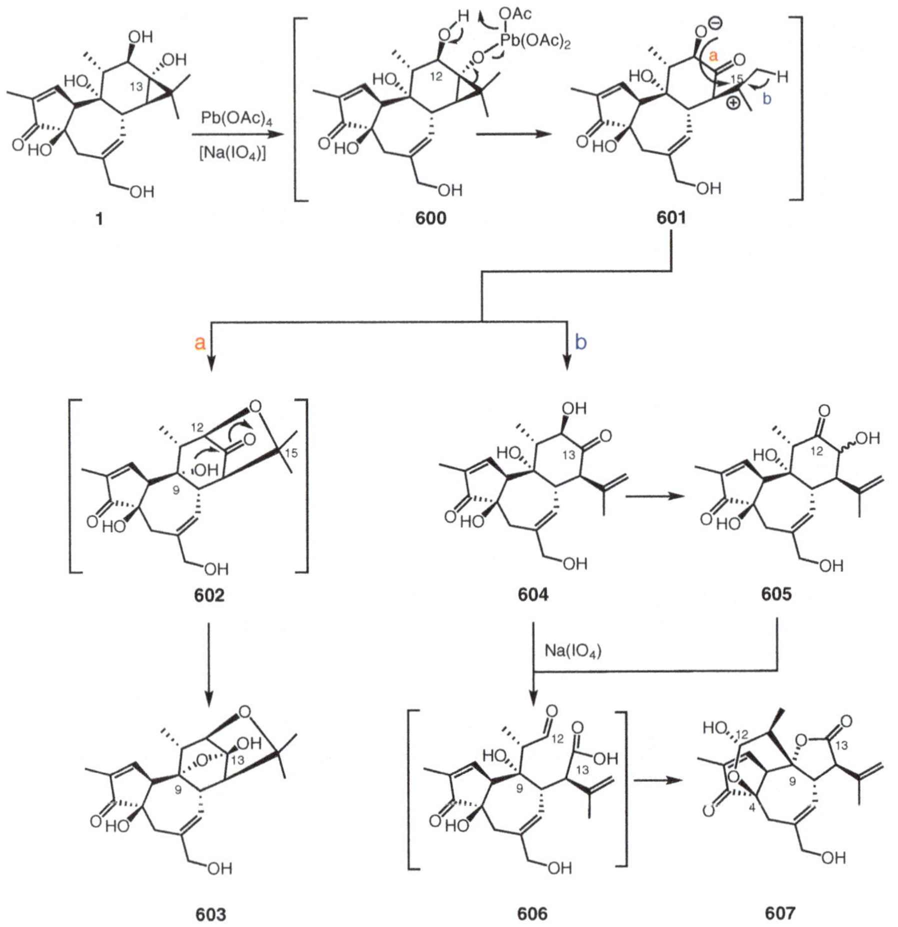

Scheme 27 Possible mechanism of the oxidation of phorbol (**1**) with lead(IV) tetraacetate and sodium periodate. All heterolytic steps could also occur in a concerted or homolytic fashion

oxidants) (Scheme 27). Heterolytic cleavage of the lead-oxygen bond could next trigger opening of the cyclopropane ring, and generation of the C-15 cation **601**. This can be trapped by the 12-hydroxy with formation of the ether **602**, and eventual formation of an hemiacetal ring between the 13-carbonyl and the 9-hydroxy group, which leads to ring C of **603** being fixed in a boat conformation. Alternatively, the C-15 cation can be quenched by proton loss from one of the geminal methyls, resulting in the formation of tigl(i)ophorbol A (bisdehydrophorbol (**604**)), from which tigl(i)ophorbol T (**605**) derives by acyloin rearrangement. With periodate, the oxidative degradation proceeds further, cleaving the α-ketol system. Both carbonyls generated in the cleavage of the C-12–C-13 bond are then engaged intramolecularly by free hydroxy groups. The C-13 carboxylate forms a γ-lactone with the tertiary hydroxy group at C-9, and the aldehyde at C-12 a δ-lactol with the C-4 hydroxy group, affording the C/D bis-*seco*-tigliane **607** [303].

The 13-hydroxy group shows a high reactivity in esterification reactions, where, at least with aliphatic acids, it is more reactive than the allylic primary 20-hydroxy group under several reaction conditions [304]. For the chemoselective synthesis of phorbol 13-acetate, the reaction was optimized with Ac_2O in DMF. The selective acetylation of the 13 and 20-hydroxy groups was possible by reaction with Ac_2O in DMF in the presence of $CaCO_3$. Cyclopropyl alcohols are more acidic than their acyclic analogues due to $n \rightarrow \sigma^*$ delocalization, which increases the electronegativity of oxygen and facilitates deprotonation [299]. However, chemoselective acylation of the C-13 hydroxy group was also reported in the absence of bases. In addition, the chemoselectivity of the mono-acylation of phorbol depends on the nature of the acid and on its activation mode. Thus, chlorides of aromatic and heteroaromatic acids react selectively on the primary 20-hydroxy group, while acetic anhydride reacts with the 13-hydroxy group [304]. Interestingly, acetylation of the 9-hydroxy group quenches the reactivity of a 13-ester in a reaction of transesterification, highlighting the relevance of intramolecular hydrogen bonding in the acylation reactivity of the 13-hydroxy group [287].

The oxygen function at C-13 shows negligible solvolytic reactivity. The ionization of the cyclopropyl substrate is associated with the disrotatory opening of the opposite cyclopropane bond and formation of a more stable allyl cation, a process that in the case of the C-14–C-15 cyclopropane ring would involve steric clash in both torquoselectivity modes allowed for the reaction.

4.4 Additional Sites of Functionalization

The Tertiary C-9 Hydroxy Group

The tertiary C-9 hydroxy group of phorbol (**1**) engages in a strong intramolecular hydrogen bonding with the C-13 ester group carbonyl. This interaction biases the orientation of the ester residue, decreasing the conformational mobility of its carbonyl head, and is of great relevance for the interaction of tigliane esters with protein kinase C (see Sect. 6.1). It also contributes to insulate, in terms of reactivity, the C-9 hydroxy group, which rarely gets involved in reactions that do not alter the basic constitution of the tigliane skeleton. One exception is the reaction of the phorbol 4,12,13,20-tetraester **608a** with thionyl chloride in pyridine, which caused dehydration of the 9-hydroxy group and formation of the dienone **609** [287]. Interestingly, the reaction of the corresponding 4-deacetyl derivative (**608b**) was accompanied by elimination of acetic acid and formation of the extensively conjugated triester **610** [287].

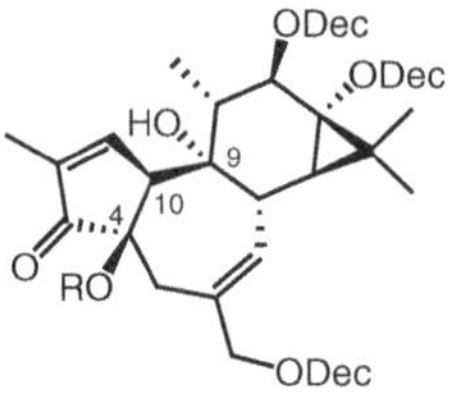

608a: R = Ac
608b: R = H

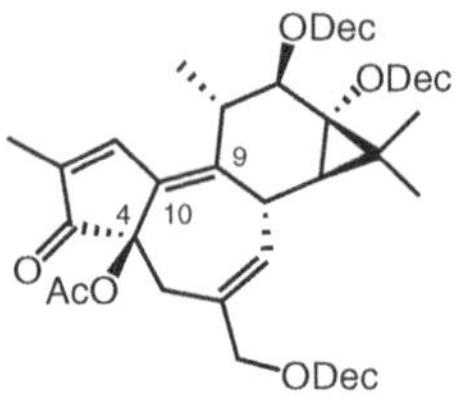

609

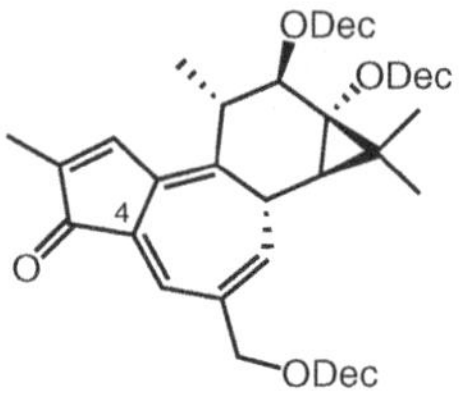

610

The C-16 Hydroxymethylene Group

Due to the instability of the the parent polyol 16-hydroxyphorbol (**12**) under the conditions of hydrolysis/transesterification of its corresponding esters, little is known about the reactivity associated with the 16-hydroxy group. An additional source of confusion is represented by the uncertaincies associated with the location of the *gem*-dimethyl hydroxylation site (see Sect. 3.4). As will be discussed in a forthcoming review on daphnanes in this book series, the 16-hydroxy group could be involved in the tigliane-daphnane rearrangement, a reaction still elusive under laboratory conditions, but conceivable in mechanistic terms.

The C-5 Oxymethine Group

The presence of a 5β-hydroxy group is generally associated to α-epoxidation of the $\Delta^{6(7)}$ double bond, generating a structural motif very rare in tiglianes but rather common in daphnanes (see Sect. 3.3). The 5β-hydroxy group easily forms a six-membered acetonide by reaction with acetone or 2,2-dimethoxypropane [8, 27]. The reaction is interesting for the synthesis of analogues of tigilanol tiglate (**3**) from the acetonide **560a**, available in two steps from the natural product or from a mixture of analogues occurring in its source plant (**560b**) [36], by reaction with dimethoxypropane and then transesterification with methanol (see Scheme 21) [33].

The 5β-hydroxy group stabilizes the A/B-ring junction toward vinylogous retro-aldol epimerization [305], possibly because of the steric hindrance it would exert in the concave face of the cup-shaped A/B-ring of the 4α-phorbol framework. Oxidation of the 5-hydroxy group of tigilanol tiglate (**3**) afforded the unstable α-hydroxy-β-dicarbonyl tigliane **611a**, which could be characterized but proved too labile for bioactivity evaluation [306]. Investigation of the degradation of its 20-silyl derivative **611b** disclosed a remarkable rearrangement of the A/B ring system, with eventual formation of an A-*seco*-δ-lactone having an oxygen bridge between C-9 and C-5 (**614**) (Scheme 28) [306]. This rearrangement bridges the reactivity of the ring A acyloin system, the oxygenated C-5 carbon, and the generally unreactive C-9

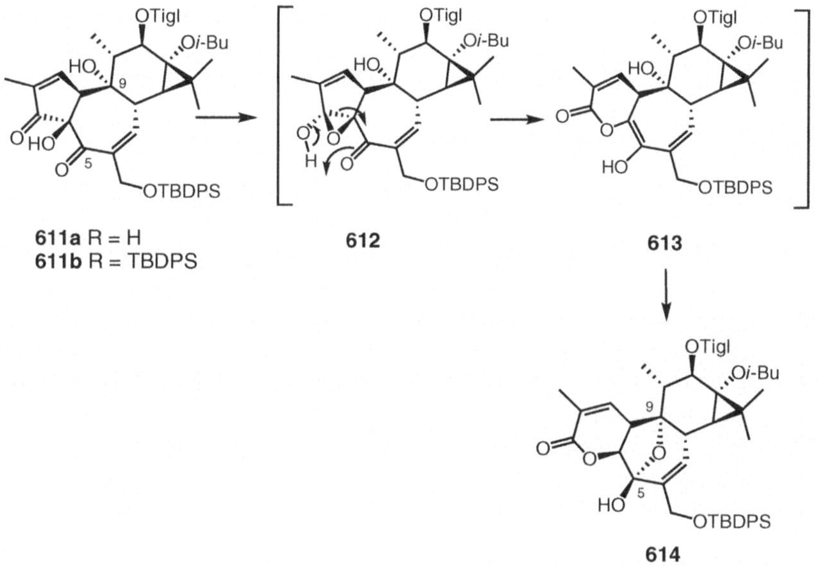

Scheme 28 Rearrangement of 20-*O*-silyl-5-dehydrotigilanol tiglate (**611b**) to the A-seco-lactone **614** (TBDPS = *tert*-butyldiphenylsilyl)

hydroxy group. Key to the rearrangement is the tautomerization of the α-hydroxy-β-dicarbonyl system of **611b** to a hydroxyepoxy intermediate (**612**), which then evolves via **613** into the *seco*-δ-lactone **614**. Only one precedent for this rearrangement seems to have been reported in the literature [307], and the reaction could be general for polycyclic α-hydroxy-β-dicarbonyls.

4.5 Summary of the Reactivity Aspects

Given its polyfunctionalization, phorbol (**1**) shows a very rich chemistry. The efforts to clarify its structure by classic degradation studies were hampered by the unusual properties and reactivity of some of its functional groups, especially the cyclopropanol element at the C/D-ring junction. Nevertheless, these studies provided a basis to unravel the reactivity of the natural polyol and support structure-activity studies. With the exception of the reduction of the 3-carbonyl and the deoxygenation of its adjacent carbon, no modification of the connectivity or configuration of the diterpenoid skeleton proved compatible with bioactivity, at least that associated with protein kinase C, but activity could be modulated by variation of the acylation profile. To guide medicinal chemistry studies, a brief recap of the reactivity of the various hydroxy groups in acylation reactions seems appropriate at the end of this section of this contribution.

(a) The hydroxy groups at C-12, C-13, and C-20 can be acylated with acids activated via anhydrides, chlorides or with carbodiimide-based coupling agents. The tertiary C-13 and the allylic primary C-20 hydroxy groups show somewhat similar reactivities, but conditions of selective mono-esterification of either hydroxy groups are possible.
(b) Acetylation of the tertiary C-4 and C-9 hydroxy groups is possible on 12,13,20-triesters using acetic anhydride and *p*-toluenesulfonic acid.
(c) In transesterification rections, the most reactive acetyl groups are those at C-4 and C-20.

Also of great relevance is the observation that bulky silylating reagents (trityl chloride, *t*-butyldimethylsilyl chloride or even more hindered reagents) react selectively with the C-20 primary allyl hydroxy group, and that etherification with the couple methyl iodide-silver oxide can only be carried out with compounds having the 13-hydroxy group acylated.

The reactivity studies also identified various incompatibilities of phorbol derivatives. In particular:

(a) aerobic conditions oxidize the hydroxylated allylic moiety on ring B;
(b) basic conditions trigger the vinylogous retro-aldol epimerization of the C-4 hydroxy group, and
(c) acidic conditions and certain transition metal based oxidants (lead tetraacetate) open the cyclopropane ring to produce mixtures of D-*seco* compounds.

Information on the instability of tigliane esters is useful for the detoxification of protein-rich oleaginous plants before their use as animal feed, and has generated protocols that combine, in addition to microbial fermentation, heat, alkaline treatments and air exposure (see Sect. 6.4).

5 Total Synthesis

Tigliane diterpenoids are highly challenging targets for total synthesis, and efforts in this area have, unsurprisingly, focused on phorbol (**1**) and prostratin (**192**), the two most well-documented and best known members of the family. A large number of ingenious entries into the tetracyclic tigliane skeleton have been developed, second perhaps in isoprenoids only to entries into the skeleton of Taxol, but only in few cases could these preliminary efforts be finalized into natural products. As a matter of fact, total synthesis has been the single most thoroughly reviewed topic in the tigliane secondary literature from the past decade [308–313]. For this reason, in an effort to limit overlap with the existing reviews, the present treatment will start in medias res, namely at the stage of pre-tiglianes, the advanced polycyclic precursors of the final targets. Conversely, reviews in this area focus mainly on the construction of these compounds, a topic focused on carbon-carbon bond formation and of more general relevance for the synthesis community compared to the idiosyncratic functional group reactivity associated with the tetracyclic tigliane skeleton. To complement information from previous reviews in the area, the ingenious ways in which pre-tiglianes have been turned into the final targets will be the leitmotiv of this section, where, for the sake of clarity, tigliane numbering will be used throughout. The synthesis challenges to convert pre-tiglianes into final natural products were not only the insertion of missing functional groups or the modification of the existing ones, but also the timing of these modifications, since the polyfunctionalization of the natural products poses problems of chemo-, regio- and stereoselectivity that need to be harmonized through a carefully planned time sequence of transformations.

The potential of synthesis efforts into the construction of the tigliane constructs goes beyond the sheer duplication of Nature, since they could be used to explore areas of the tigliane pharmacophore inaccessible by isolation or semisynthesis. Unfortunately, in the current scenario, the focus of natural product synthesis seems to be on the duplication of Nature, something that only rarely competes with the isolation work, and the laurel goes to the person who first plants his/her flag on top of the mountain rather than to those who use the opened pathway to explore the mountain. As a result, the huge synthesis undertaking on tiglianes has remained substantially underdeveloped from a biological standpoint, although, at least in Wender's case, it provided expertise later applied to increase, by semisynthesis from phorbol, the availability of prostratin (**192**) and tigilanol tiglate (**3**), two tiglianes of biomedical relevance, and, at least for prostratin, in short supply by isolation (see Sect. 6).

Comparison between the various syntheses in terms of efficiency is complicated for several reasons, with one being the evolution of what is meant by a step. The definition of step is generally considered nowadays as independent from the number of transformations involved in a single laboratory operation, and refers to the conversion of a substrate to a product in a single reaction flask, without intermediate work-ups and/or purifications [314]. In the past few years, the concept of a step was, conversely, often related mainly to single molecular transformations. Since experimental details have not always been disclosed in the original publications, the relative advantage of telescoping reactions into single synthesis steps vs. an increase of overall yield for a set of independent steps could not be clearly defined. In any case, the arithmetic of step-counting is an imperfect rating system for synthetic practicability, overlooking important factors like scalability and the nature of the starting material, both absent in the step-counting meritocracy scale. Thus, the concept of "commercial availability" has a monetary-, a temporal-, as well as a geographical-component. A synthesis with a cheap starting material of general availability surely has merits over one that uses an exotic and expensive compound. In addition, despite globalization, a chemical can have a local availability or its trade could be complicated due to legal restrictions. A synthesis starting from safrole should surely raise concerns because of the legal status of this compound, the starting material for the synthesis of the recreational drug 3,4-methylenedioxymethamphetamine (MDMA, ecstasy) [315].

Rhamnopholane (13,15-*seco*-tiglianes)- and *nor*-rhamnopholane intermediates are involved in most syntheses of tiglianes, but the total synthesis of diterpenoids of this class will be discussed in a forthcoming review on ring D *seco*-tiglianes (daphnanes, rhamnopholanes). In addition, the synthesis of rhamnopholanes eludes the most peculiar step of tigliane synthesis, namely the generation of the hydroxylated C/D ring junction.

5.1 *The Wender Syntheses of Phorbol (1989–1997)*

The first total synthesis of phorbol (**1**), and of a tigliane diterpenoid in general, was accomplished by Wender at Stanford University in 1989 [297]. Wender then refined his racemate synthesis, a 52-step "tour-de-force", in further-generation syntheses, addressing the issues of efficiency (second-generation synthesis) [316], and then of chirality (third-generation synthesis) [317]. The three syntheses all rely on oxidopyrilium-alkene [5+2] cycloadducts (Scheme 29, **A**) [318] to generate tricyclic proto-tigliane systems (**B**), next converted into pre-tigliane tetracyclic systems (**C**) by the generation of ring A through the agency of an intramolecular dipolar cycloaddition (first generation synthesis) or of a transition metal-promoted enyne cyclization (second- and third generation syntheses) (Scheme 29).

The 1989 first-generation synthesis [297] started from furfuryl alcohol (**615**), converted into the advanced prototigliane **616** in 11 steps (Scheme 30) [318]. The formation of ring A capitalized on the stepwise conversion of the C-4 cyano group of **616** into a nitroxide, which spontaneously underwent [3+2] dipolar cycloaddition to the isoxazoline **617**, and was then converted, in uneventful six steps, into the pretigliane **618**. The functional group decoration of **618** next was modulated by hydrogenation of the $\Delta^{2(19)}$ double bond, reductive cleavage (DIBAL-H) of the benzoate protection of the 12-hydroxy group, and oxidation to the 12-keto derivative **619**. The introduction of the cyclopropane D-ring required conversion of the saturated 12-ketone into the $\Delta^{12(13)}$ α-acetoxyenone, a maneuver carried out in three passages (kinetic deprotonation and α-sulfanylation with phenylsulfenyl chloride, C-H oxidative functionalization at C-13 with lead tetraacetate, and oxidative elimination of the sulfanyl group by oxidation with *m*CPBA and heating with triethylphosphite), with the whole set of transformations summarized into only two operational steps. The enolate formation step was critical in terms of the regiochemistry, since enolization

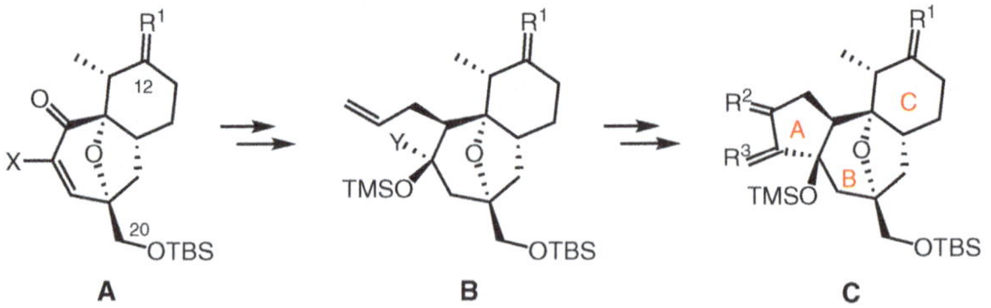

A **B** **C**

Scheme 29 Generation of pre-tigliane precursors (**C**) from prototiglianes (**B**) in turn derived from oxidopyrilium-alkene cycloadducts (**A**) in the three phorbol syntheses by Wender. R^1: CH_2=or OAc, H; X: H or OTMS; Y: CN or R-CH≡C-; R^2: CH_2= or CH_3, H; R_3: =O or RCH=)

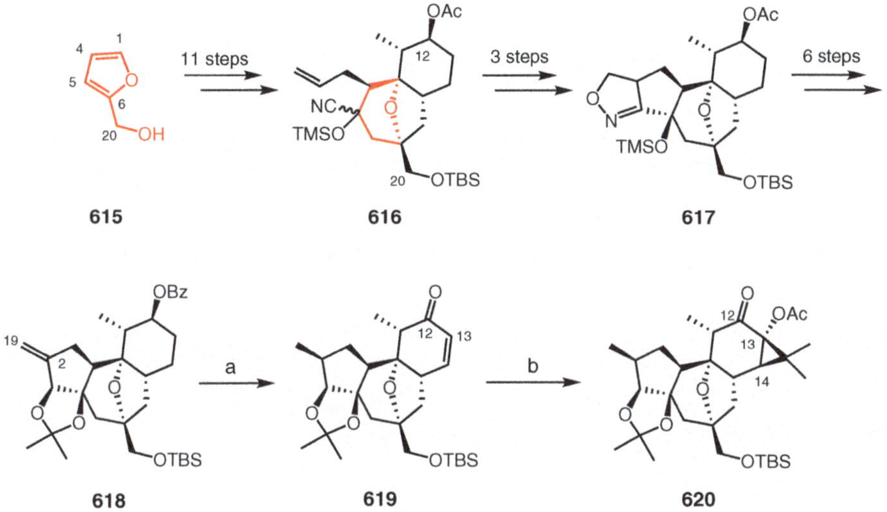

Scheme 30 First-generation Wender synthesis of phorbol (**1**). A: preparation of the tigliane ketone **620**. (a) (i) H_2, $RhCl(PPh_3)_3$; 92%. (ii) DIBAL-H; (iii) PCC (ii + iii: 94%); (iv) LDA, TMSCl; (v) PhSCl (iv + v … 77%); (vi) $Pb(OAc)_4$ (78%); (vii) *m*CPBA; viii) Δ, $P(OEt)_3$. (b) $Ph_2S{=}C(CH_3)_2$ (vii + viii … 85%). DIBAL-H = diisobutylaluminum hydride; LDA = lithium diisopropylamide; *m*CPBA = *m*-chloroperbenzoic acid; PCC = pyridinium chlorochromate; TMSCl = trimethylsilyl chloride

on the methine site (C-11) of the carbonyl could have jeopardized the configuration of this stereocenter. The dimethylcyclopropane element was introduced by addition of its corresponding dimethylsulfonium ylide ($Ph_2S{=}CMe_2$), for which its agency generated stereoselectively the tigliane ketone **620** (Scheme 30).

Being able to obtain the natural configuration of the C-12 hydroxy group turned out to be a path strewn with unexpected complications, since the borohydride reduction of **620** afforded the incorrect 12α-isomer. Steric congestion associated with the adjacent and β-oriented dimethylcyclopropane ring, was instead expected to direct hydride attack from the opposite α-face, with attainment of the correct β-configuration of the C-12 hydroxy group. It seems therefore that the α-oriented C-11 methyl group governs the stereochemistry of attack to the 12-carbonyl. The formation of the 12β-hydroxy group required formation of the α-*cis*-12,13 carbonate **621** by treatment of **620** with DIBAL-H to remove the 13-acetyl, and then with carbonyl diimidazole to produce the cyclic carbonate. Opening of the carbonate was associated deftly with the remodeling of ring B. Towards this end, the 20-TBS group was removed by fluoridolysis and replaced, via the corresponding triflate, with an iodine atom. Treatment with *t*-BuLi opened regioselectively the carbonate, generating a 13-pivalate, also triggering the deiodinative opening of the 6,9-oxygen bridge, affording a $\Delta^6(20)$ tigliane alcohol, which was oxidized to the 12-ketopivalate **622** and finally transformed by means of $NaBH(OAc)_3$, DIBAL-H, and Bz_2O, DMAP into **623** (Scheme 31).

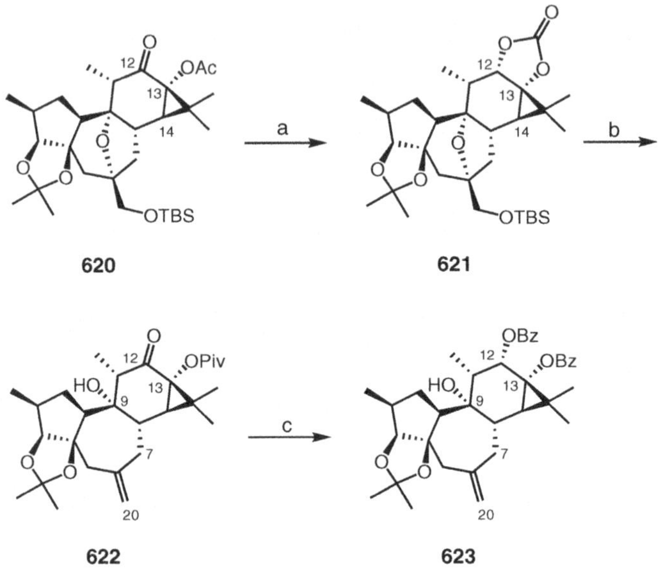

Scheme 31 First-generation Wender synthesis of phorbol (**1**). B: Establishment of the configuration of the 12,13-glycol system and preparation of the dibenzoate **623**. (a) (i) DIBAL-H; (ii) $CO(Im)_2$. (b) (i) TBAF; (ii) Tf_2O, TEA; (iii) Bu_4NI, HMPA; (iv) *t*-BuLi (45% from **567**); (v) PCC. (c) (i) $NaBH(OAc)_3$ (92%); (ii) DIBAL-H; (iii) Bz_2O, DMAP. DMAP = 4-dimethylamino pyridine; HMPA = hexamethylphosphoramide; TBAF = tetrabutylammonium fluoride; TEA = trimethylamine

A semi-synthesis version of the pivalate **622** was then obtained from phorbol triacetate (**552**) and used to complete, in a relay fashion, the synthesis. The chemistry involved has been discussed in Sect. 4, and is summarized in Scheme 32 (stereoselective reduction of the C-3 carbonyl, acetonide formation between the adjacent hydroxy groups at C-3 and C-4, and chemoselective removal of the 20-acetate to afford **624**; oxidation of the primary 20-hydroxy group to a 20-aldehyde to avoid allylic deoxygenation during the catalytic reduction, elimination of the 20-hydroxy group to afford the 20-exomethylene tigliane **625**; global deacetylation and chemoselective pivaloylation of the 13-hydroxy group).

The synthesis continued with the stereoselective reduction of the 12-carbonyl to a 12β-alcohol by chelate reduction with $NaBH_3CN$. This poorly reactive borohydride only reduces α-hydroxy- and α-acyloxyketones via internal hydride delivery, securing formation of the counter-kinetic course of the reduction by hydride attack from the α-face of the carbonyl, where the adjacent oxygen function is located. The 13-pivalate **622** was then reductively cleaved with DIBAL-H, and the 12,13-glycol system was protected by treatment with benzoic anhydride, eventually affording the dibenzoate **623** (Scheme 31).

Having passed the bottleneck of the functionalization of rings C and D, the synthesis was completed by the remodulation of the functionalization of rings A

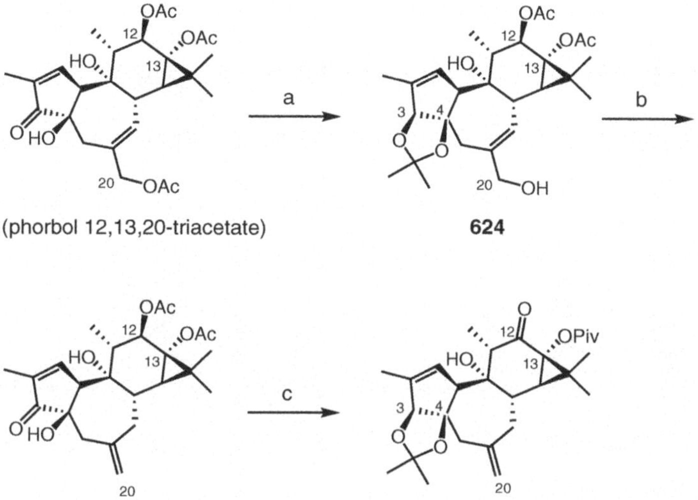

552 (phorbol 12,13,20-triacetate) **624**

625 **622**

Scheme 32 Semi-synthesis of the ketopivalate **622** from phorbol 12,13,20-triacetate (**552**). (a) (i) $NaBH_4$, $CeCl_3$; (ii) DMP, PPTS; (iii) $HClO_4$, MeOH. (b) (i) MnO_2; (ii) H_2, Pd(C); (iii) ArSeCN, (n-Bu)$_3$P, mCPBA. (c) (i) KCN, MeOH; (ii) Me_3COCl, DMAP. DMP = 2,2-dimethoxypropane; PPTS: pyridinium p-toluensulfonate

and B. Thus, allylic oxidation of the ring B exomethylene double bond of **623** with SeO_2 generated the 7β-hydroxy derivative **624**, and allylic displacement with thionyl chloride provided an endocyclic allylic C-20 chloride, with the halide being eventually displaced with silver benzoate to afford the tribenzoate **625** (Scheme 33). The elaboration of ring A involved hydrolysis of the 3,4-acetonide function, oxidation of the C-3 allyl alcohol and protection of the tertiary 4-hydroxy as a TMS ether to avoid its isomerization during formation of the Δ^1 double bond by α-bromination to the C-3 carbonyl. Base-induced elimination afforded the phorbol derivative **626**. Demasking of the hydroxy functions first was carried out by treatment with PTSA in methanol for the 4-silyl group, and then with methanolic KCN for the three benzoate groups, eventually affording phorbol (**1**) in 52 steps from furfuryl alcohol (**615**), and in an overall 0.23% yield.

The second-generation phorbol synthesis by Wender [316] features a streamlined preparation of two precursors of the acetoxyenone **619**, the ring C ketones **627a** and **627b** having latent carbonyls at C-3 and C-12 (Scheme 34). The generation of ring A was improved by replacing the nitrile oxide cycloaddition with an enyne intramolecular cyclization of substrates (**627a** and **627b**) having C-3 in what will become a masking ethylidene group for the C-3 carbonyl (Scheme 33). In this way, it was possible to prevent the easy isomerization of the 4-hydroxy group under the basic conditions required for the installation of the ring C acetoxyenone system, a prelude to the cyclopropanation step [297]. The enyne cyclization of **627a** to **628a** was carried out reductively under palladium catalysis [(Pd_2(dba)$_3$] and in the presence

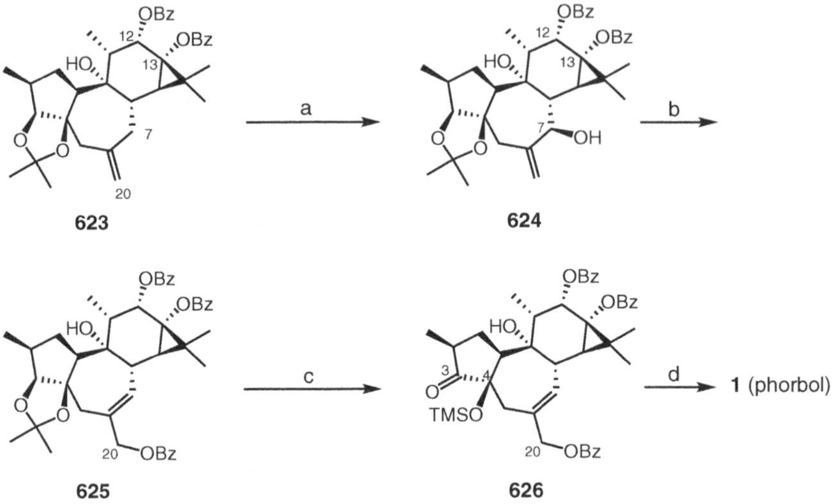

Scheme 33 First-generation Wender synthesis of phorbol (**1**). C: Completion of the synthesis by establishing the functionality of rings A and B. (a) SeO_2, TBHP (50%). (b) (i) $SOCl_2$ (80%); (ii) AgOBz, KOBz (73%). (c) (i) $HClO_4$, MeOH, montmorillonite clay; (ii) SO_3-pyridine, TEA, DMSO (80%); (iii) $CF_3CON(CH_3)TMS$, DMAP (91%); (iv) $KN(TMS)_2$, TMSCl, then NBS (65%); (v) $LiBr/Li_2CO_3$, Δ (72%). (d) (i) PTSA, MeOH; (ii) KCN, MeOH (i + ii = 86%). NBS = *N*-bromosuccinimide; PTSA = *p*-toluensulfonic acid; TBHP = *t*-butyl hydroperoxide

of tri-*o*-tolylphosphine as ligand (58% yield), dimethylhydrosiloxane as a hydride donor, and acetic acid as a proteolytic agent [319]. A similar yield was observed when the enyne cyclization was carried out on the 4-*O*-silyl derivative **627b** by treatment with dibutylzirconocene, which generated a ziconacyclopentene intermediate that was then protolyzed with acetic acid [320] to the tricyclic diene **628b**. Although uninfluential for the further course of the synthesis, the stereogenic element at C-2 and at C-3 were installed with notable stereoselectivity (*β*-configuration for the C-2 methyl, (*Z*)-configuration for the double bond centered on C-3) [316]. Further ozonolytic elaboration of the cycloadducts **628a** and **628b** afforded, after reductive work-up and acetalization, **629**, which was then funneled into the first-generation synthesis scheme by conversion to the acetoxyenone **619**.

Ring A closure was further improved in the third-generation, asymmetric synthesis, as developed by Wender in 1997 [317]. The tricyclic pre-tigliane **630** was assembled in enantiomerically pure form in 14 steps from furfuryl alcohol (**615**), using the Evans aldol reaction to install chirality. Ring A was then generated using a zirconium-catalyzed enyne cyclization [320], which not only produced a zirconacyclopentene intermediate, but also deprotected the 12-hydroxy group. In this way, protonolysis generated the 3-benzylidene *nor*-rhamnopholane derivative **631** (Scheme 35). The conversion of **631** into (+)-phorbol (**1**) was then done according to the first-generation racemic synthesis, with the obvious modifications associated with the ozonolytic generation of the 3-carbonyl, and this required 19 additional

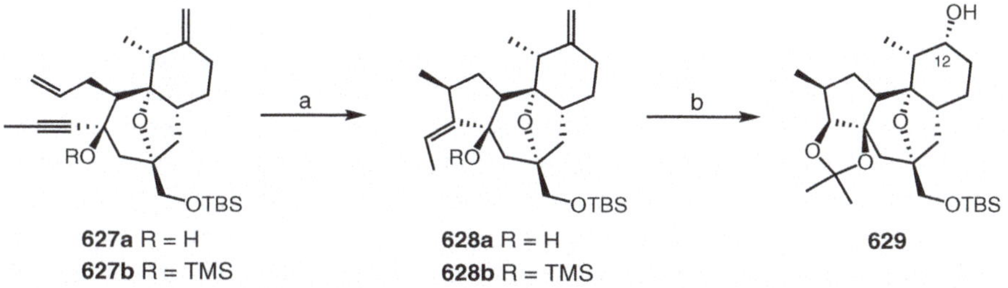

Scheme 34 Second-generation Wender synthesis of phorbol (**1**). Generation of the acetoxyenone precursors **628a** and **628b** by enyne cyclization and oxidative remodeling. (a) for **628a**: $Pd_2(dba)_3$, tri(*o*-tolyl)phosphine, HOAc, $[(CH_3)_2SiH)_2O$ (58%); for **628b**: Cp_2ZrBu_2, then HOAc quenching (69%). (b) (i) O_3, CH_2Cl_2, MeOH, $NaBH_4$ work-up; (ii) DMP, PPTS; PPC, NaOAc (overall 24% **629** after PCC oxidation of the 12-hydroxy group)

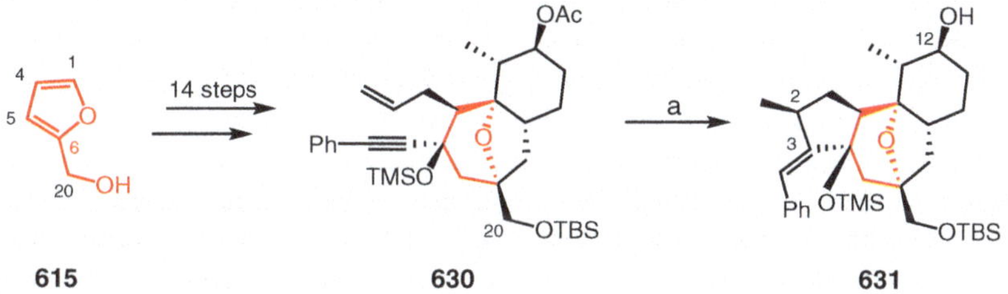

Scheme 35 Third-generation Wender synthesis of phorbol (**1**). Generation of the tricyclic C-3 benzylidene norrhamnopholane **631** by enyne cyclization. a) Cp_2ZrCl_2, *n*-BuLi, HOAc (93%)

steps. The overall linear step count was 35, but homochirality came at no price in terms of step-counting, and the overall yield (0.11%) compared to the first- and the second-generation syntheses.

A tricyclic pre-tigliane intermediate analogous to **631** was prepared in 2001 by Cha using a different strategy from that of Wender, replacing zirconium with titanium to promote the enyne cyclization step [321].

5.2 *The Baran Synthesis of (+)-Phorbol (2016)*

This synthesis was a chiral-pool-based enantioselective entry into phorbol (**1**), and is inspired by the biogenetic logic of the two-phase terpenoid synthesis, being focused on the building of a polycyclic system (phase one) followed by its decoration with oxygen functions (phase two) [322]. The synthesis is based on the elaboration of the (+)-3-carene (**632**)-derived tetracyclic 19-nortigliane **633**, lacking, in addition to the C-2 methyl, also functionalization at C-3, C-12, C-13, and C-20, and bearing a redundant oxygen function at C-7 as well as a surrogate olefin functionalization for the C-4 and C-10 stereocenters [322] (Scheme 36). The 19-nortigliane **633** was assembled from **632** in six steps, and had been used previously for a total synthesis of ingenol [323]. Mukaiyama hydration [324] of the $\Delta^{1(10),4}$-dienone system occurred on the terminal double bond and with Markovnikov regiochemistry, overcoming the bias for addition at the terminal carbon observed in β-substituted α,β-unsaturated esters [325]. Cobalt-hydride attack [326, 327] occurred from the *β*-face of the dienone, presumably due to the presence of an *α*-methyl group on the allylic C-6 carbon. The C-4 hydroxy group was trapped in situ as a trimethylsilyl (TMS) ether, providing **634**, the substrate for the challenging C-H functionalization at C-12. Here the issue was the distinction between methine and methylene carbons, and within these, between C-12, C-3, and C-5. Electronic considerations (higher s-character and stronger C-H bond) can be advocated for the deactivation of the cyclopropane methines C-13 and C-14 [327], while the four remaining methines are sterically hindered, overall suggesting a chemoselective methylene- rather than methine-oxidation [328]. The three methylenes resonated at distinct δ values, with that of C-12 being the more upfield and therefore the more electron-rich and nucleophilic, a feature associated with the presence of an adjacent and hyperconjugating cyclopropane ring [328]. This is expected to transfer electronic density into the *β*-oriented *pseudo-equatorial* antibonding carbon-hydrogen bond, facilitating therefore its cleavage during the oxidation. In the event, treatment with trifluoromethyldioxirane afforded the 12*β*-alcohol **635** with an intact cyclopropane ring, a somewhat surprising observation in light of the radical mechanism of dioxirane C-H oxidation [329], and presumably related to an oxygen rebound step faster than "radical clock" cyclopropane opening. The cyclopropane ring was opened via a Lewis acid-catalyzed cyclopropylogous β-elimination, which afforded the rhamnopholane diene **636**. The exomethylene bond was subjected to another Mukaiyiama hydration reaction, carried out in the presence of triphenylphosphine to quench the reactive peroxidic radical generated in the reaction [326], affording the Δ^{12}-olefin **637** in 34% yield, with the remaining being mostly starting material **634**, recovered at the significant extent of 62% from the telescoped reaction sequence (Scheme 36).

The stage was now set for the critical step of the synthesis, namely, the combined introduction of oxygen atoms at C-12 and C-13 and the closure of the cyclopropane ring. The disubstituted double bond of **637** was first converted to the corresponding α-diketone **638**, a reaction requiring considerable effort (Scheme 37). After much experimentation, an almost quantitative yield nevertheless could be obtained with the high-valence ruthenium species generated in situ from $RuCl_3$ and $NaBrO_3$. Perruthenate generation with oxone or permanganate under a variety of conditions either failed or did not give reproducible results, as did the replacement of ruthenium trichloride with osmium trichloride [322]. A reductive re-installation of the cyclopropane ring of the Δ^1-carene starting material was planned via the formal attack of a nucleophilic 12,13-ene-diol to the electrophilic central carbon of the adjacent isopropyl group. The strategy developed involved: (a) activation of the side chain C-15 hydroxy group to **639**, (b) two-electron reduction of the α-diketone group to an ene-diol, and (c) intramolecular cyclization, but the desired product **643** was obtained with a rather roundabout mechanistic, yet high-yielding, course. Thus, treatment of the tertiary

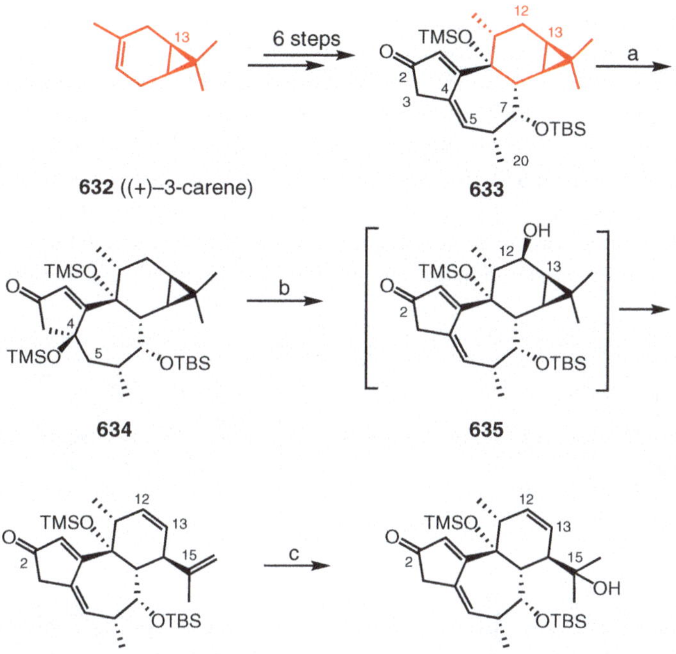

Scheme 36 Baran synthesis of phorbol (**1**). A: from Δ^1-carene (**632**) to the 19-norrhamnopholane alcohol **637**. (a) $Mn(acac)_2$, $PhSiH_3$, O_2, then TMSOTf, Et_3N, quench (70%). (b) TFDO, then ZnI_2, MgI_2, Et_2O. (c) $Mn(acac)_2$, $PhSiH_3$, O_2, PPh_3 (34% from **634** + 62% recovery of **634**). TFDO = trifluoromethyldioxirane

trifluoroacetate **639** with zinc triggered a formal intramolecular pinacol coupling between the carbonyl groups of the 13-ketone and that of the ester trifluoroacetate, affording the crystalline semiacetal **640**, for which the structure was confirmed by X-ray crystallography [322]. Rather than by direct diradical coupling, the semiacetal **640** could be better derived from a post-reductive aldol-type nucleophilic attack of the dienol to the trifluoroacetate carbonyl, a sp^2-electrophilic site for which it evidently has more affinity than the sp^3-electrophilic site at C-15 of the side chain. Whatever the mechanism, the synthesis was at an apparent dead end, but eventually overcome in an ingenious manner. Thus, treatment with Ac_2O trapped the vicinal C-13 hydroxy and the semiacetal hydroxy as the semi-orthoester **641**, which, upon treatment with triethylamine, underwent reorganization of the furanoid system to the desired 12-acetyloxy tigliane **642**. Direct generation of the enediol by treatment of the formal pinacol coupling product with the Lewis acid $Sc(OTf)_3$ led to rearrangement to a phorbolbutanone derivative, as expected from the study of the reactivity of 12-dehydrophorbol with Lewis acids (see Sect. 4.3). Overall, the generation of the 13-acetyloxynortigliane **643** from the *nor*-rhamnopholane α-diketone **638** involved thirteen discrete bond-forming and bond-cleaving events, and was carried out on a one-gram scale in a single, telescoped one-pot step.

Conjugate reduction of the $\Delta^{1(10)}$-enone bond of **643** represented then a considerable challenge, and the C-10 stereocenter was eventually was introduced by $NaBH_3CN$ reduction of the tosylhydrazone of the carbonyl at C-2 (Scheme 38). With a 40% yield of **644** from **643**, this was the nadir of the synthesis in terms of efficiency. The introduction of the C-3 carbonyl oxygen was then carried out with the couple CrO_3/3,5-dimethylpyrazole, generating the 3-keto-19-nortigliane **645**. Allylic oxidation was selective for the C-3 methylene vs. the C-10 methine, as generally observed

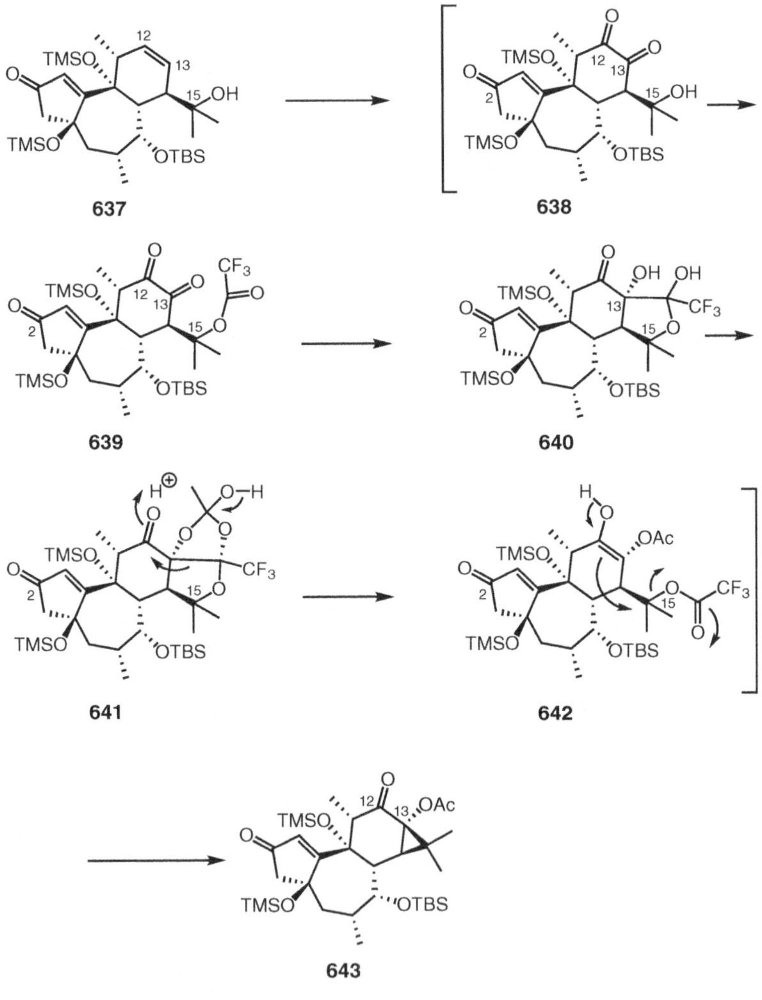

Scheme 37 Baran synthesis of phorbol (**1**). B: Conversion of the *nor*-rhamnopholane **638** into the nortigliane **643**. (a): $RuCl_3$, $NaBrO_3$ (96%). (b): CF_3COOH, DMAP, then Zn, AcOH, then Ac_2O, DMAP, then Et_3N (overall 64%)

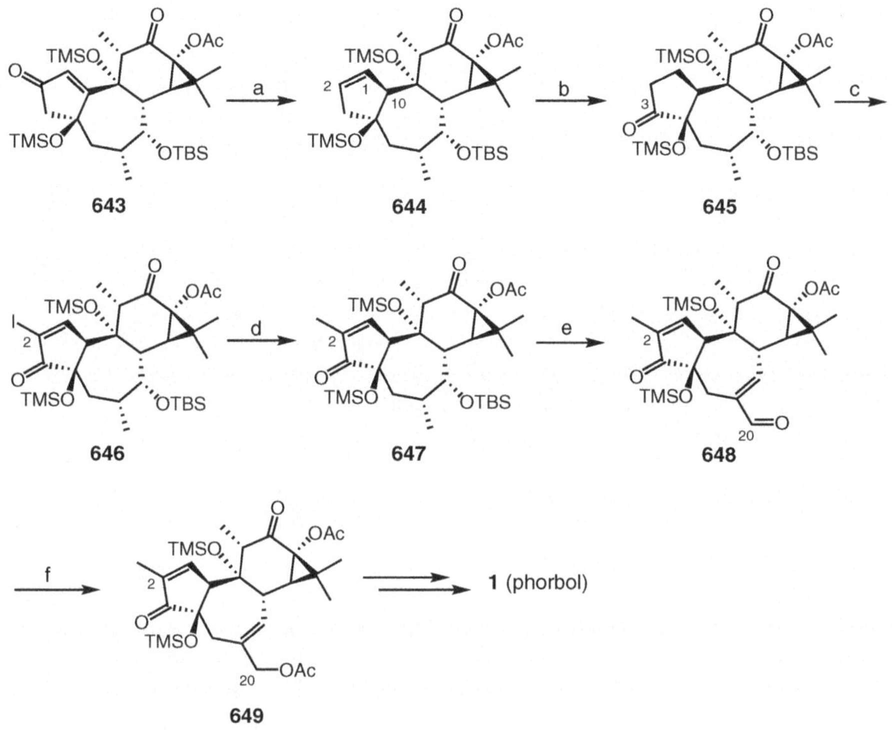

Scheme 38 Baran synthesis of phorbol (**1**). C: Completion of the synthesis by adjustment of the rings A and B functionalization. (a) $TsNHNH_2$, MeOH, Δ, then $NaBH_3CN$, AcOH, Δ (40%). (b) CrO_3, 3,5-DMP (46%). (c) $TMSN_3$, then I_2. (d): Me_4Sn, $PdCl_2(PhCN)_2$, $AsPh_3$, CuI, NMP (64% from **545**). (e) (i) HF·pyridine; (ii) Martin sulfurane, then SeO_2 (51% from **545**). (f) $NaBH_4$, then Ac_2O, DMAP (93%)

around non-symmetrically substituted double bonds. The stage was now set for the introduction of the C-19 methyl, carried out in two steps. The first was treatment with TMS-N_3 and iodine, which afforded the 2-iodo derivative **646** as a result of conjugated addition of TMS-N_3, enolate trapping by iodine and β-elimination of hydrazoic acid [330]. The iodine atom at C-2 was then changed into a methyl by Stille coupling with Me_4Sn and $PdCl_2(PhCN)_2$ in the presence of triphenylarsine and CuI [331], completing with **647** the carbon inventory of the tigliane skeleton. The following step was the selective deprotection of the trimethylsilyl (TMS) group on the C-9 hydroxy group and of the *t*-butyldimethylsilyl (TBS) group on the C-7 hydroxy with HF-pyridine. Unexpectedly, the TMS group on the tertiary C-4 hydroxy remained unscathed. Treatment with the Martin sulfurane oxyphilic reagent triggered water elimination and formation of the Δ^6-olefin, and the C-20 methyl could then be oxidized successfully with SeO_2 in a telescoped reaction carried out in the same flask of the dehydration, affording the 20-aldehyde **648**. The synthesis was completed by reduction of the carbonyl functions at C-20 and at C-12, which was carried out sequentially with, respectively, $NaBH_4$ and $NaBH(OAc)_3$. Two independent steps were necessary since the $NaBH_4$ reduction of the 11-ketone affords stereoselectively the unnatural 12α-hydroxy derivative, as discussed in the section on the Wender synthesis. Interestingly, the 20-hydroxy group had to be protected as an acetate before the chelation-controlled reduction of the 11-carbonyl to avoid reduction of the 3-carbonyl. The long-range effect of acetylation of the C-20 hydroxy group on the reduction of the C-3 enone is surprising, and highlights how relatively small structural perturbations can have marked effects not only on bioactivity, but also on reactivity, with the 3-carbonyl apparently "feeling" what goes on on the 20-hydroxy group. Global deprotection with TBAF for the 4-silyl group and $Ba(OH)_2$ for the 20-acetate **649** eventually afforded phorbol (**1**) in nineteen steps and an overall 0.051% yield from (+)-carene (**632**), a considerable improvement in step count, since in all previous syntheses the number of steps was >30.

5.3 The Li Synthesis of (±)-Prostratin (2018)

This synthesis uses a highly functionalized advanced intermediate, the bicyclic B-*seco*,D-*seco*-enone **651**, available in ten steps from *p*-methoxyphenylmagnesium bromide and (±)-*cis*-5-(hydroxymethyl)cyclopent-2-enol (**650**), in turn prepared from commercial cyclopentadiene (Scheme 39) [332, 333]. The acetoxylated cyclopropane ring was introduced according to the Wender protocol (formation of a $\Delta^{1(5)}$-pyrazoline, oxidative isomerization to an acetoxy $\Delta^{1(2)}$-pyrazoline by lead tetraacetate, photolysis) [295]. This protocol affords a single isomer when carried out on tricyclic rhamnopholane substrates, but the higher conformational flexibility of **651** led to the formation of two diastereomers in a ca. 8:1 ratio. The major one was that with the correct configuration of the cyclopropane junction, and was desilylated during work up, with formation of **652**. The secondary 4-hydroxy group of the major *cis*-cyclopropahexane isomer was oxidized, and then treated with 2-methylallylzinc chloride. The attack of the organometallic reagent occurred from the less hindered α-face of the carbonyl, affording the tertiary alcohol **653**, having the correct configuration of the 4-hydroxy group. Ring-closing metathesis with a second-generation Grubbs catalyst generated the tigliane derivative **654**, for which the 3,19-acetonide protection was removed by treatment with catalytic amounts of DDQ in a ternary solvent mixture (acetonitrile/THF/H_2O, 8:2:1). This somewhat unusual acetal deprotection was followed by the selective pivaloylation of the primary C-19 hydroxy group, a step necessary to secure β-elimination and formation of a $\Delta^{3(19)}$ double bond in the oxidation following that of the C-3 hydroxy with 2-aza-adamantane *N*-oxyl (AZADO), to afford the exocyclic olefin **655**. After silylation of the tertiary 4-hydroxy group, allylic oxidation of the 20-methyl with SeO_2 was followed by isomerization of the exocyclic $\Delta^{3(19)}$ bond to the endocyclic $\Delta^{1,2}$ location with $RhCl_3$, affording **656**. Chemoselective reduction of the 20-aldehyde group with $NaBH_4$ eventually afforded (±)-prostratin (**192**). The synthesis required 20 steps, with an overall yield of 0.34%.

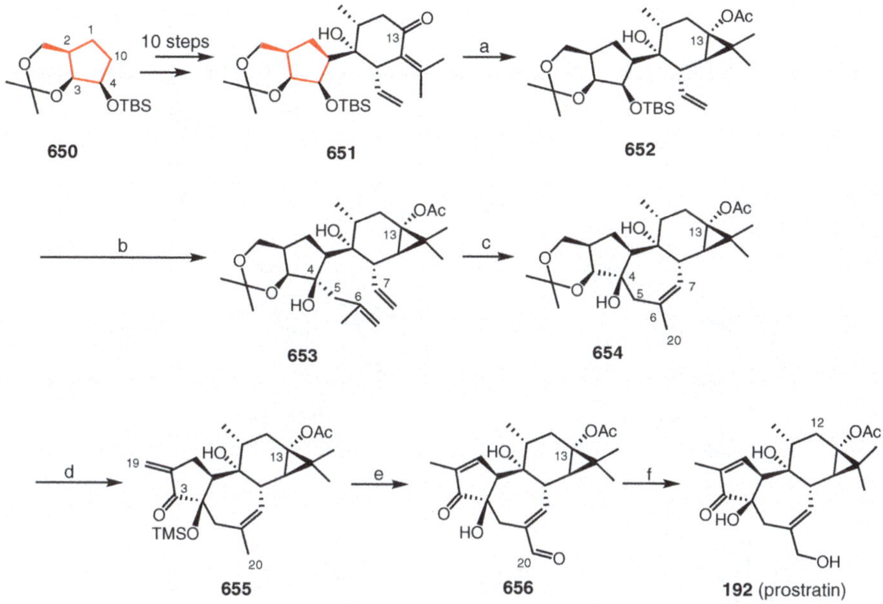

Scheme 39 Li synthesis of (±)-prostratin (**192**) (shown for an arbitrary enantiomer of the racemate). (a) (i) N_2H_2, TFA, Δ; then $Pb(OAc)_4$, (52%); (ii) *hν*, then TBAF (79%). (b) (i) TPAP, NMO, 4 Å MS; (ii) $ZnCl_2$, 2-methyl-2-propenyl magnesium chloride (i + ii … 62%). (c) Grubbs second-generation catalyst, Δ (71%). (d) (i) DDQ, then PivCl, pyridine, DMAP (73%); (ii) AZADO, DMAP, CuCl, 2,2′-bpy; TMSCl, imidazole, (64%). (e) (i) SeO_2 (81%); (ii) $RhCl_3·3H_2O$, EtOH, Δ (52%). (f) $NaBH_4$, MeOH (86%). AZADO: 2-aza-adamantane *N*-oxyl; 2,2′-bpy = 2,2′-dipyridine; DDQ = dichlorodicyanobenzoquinone; TBAF: tetrabutylammonium fluoride

5.4 The Inoue Synthesis of (+)-Prostratin (2021)

Just as the 2016 Baran synthesis of phorbol (**1**) was part of the development of the more general and bioinspired two-phase logic of terpenoid synthesis, so that of (+)-prostratin (**192**) by Inoue (Scheme 40) was developed in the framework of establishing a general entry into various phorboid families starting from the enantiopure tetra-secotigliane **658** [334]. This was prepared in ten steps from 3-hydroxy-2*H*-pyran-2-one (**657**), incidentally also the starting material for the Nicolaou synthesis of Taxol [335]. The presence of a C-3 carbonyl assisted in the introduction of the 19-methyl as an exomethylene by reaction with the Eschenmoser salt $[H_2C=N(Me)_2]^+$. Isomerization of the C-3 exomethylene of **659** to the endocyclic position was effected by treatment with the polynuclear ruthenium complex $Ru_3(CO)_{12}$, followed by telescoped treatment with KO*t*-But, molecular oxygen and triethylphosphite to afford a mixture of hydroxylated C-4 epimers, with the 4*β*-epimer **660** prevailing over the unwanted 4*α*-epimer by a modest 1.7:1 ratio. The unwanted epimer could be partially recycled by deoxygenation with SmI_2 and re-hydroxylation, and the synthesis continued with the acid-promoted methanolysis of the TIPS-protected

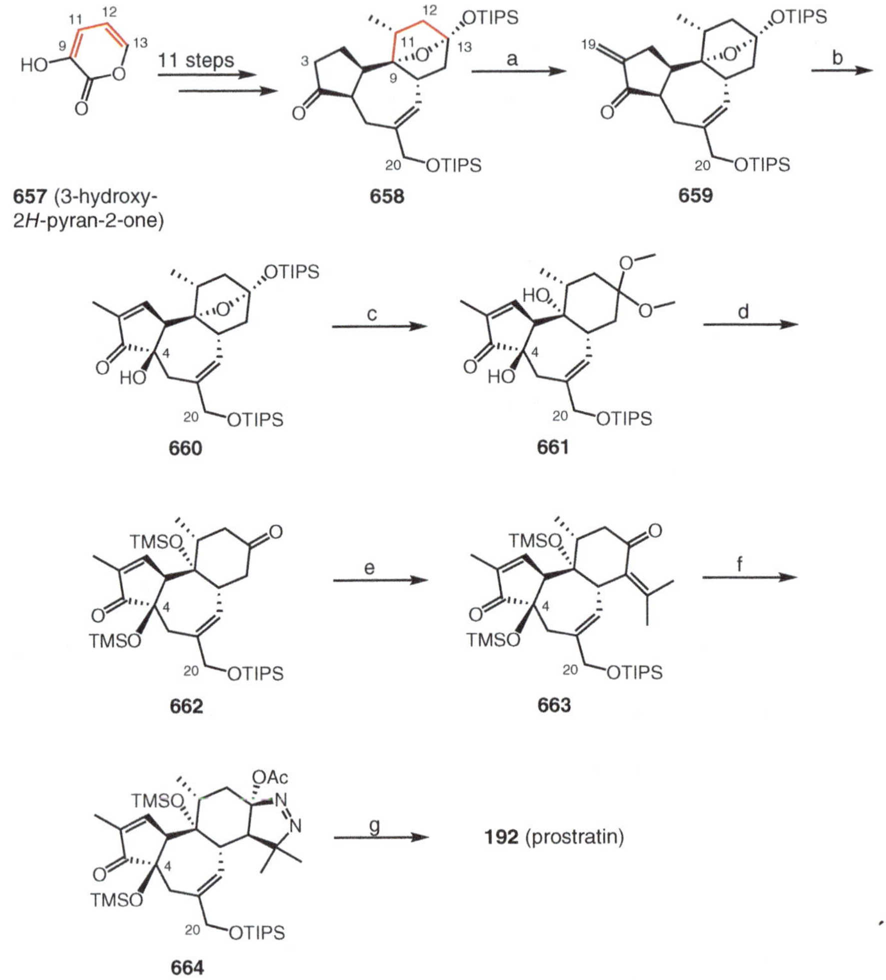

Scheme 40 Inoue synthesis of (+)-prostratin (**192**). (a) $LiN(TMS)_2$, then $[H_2C{=}NMe_2]I$ (76%). (b) (i) $Ru_3(CO)_{12}$, TEA, Δ; (ii) *t*-BuOK, O_2, $P(OEt)_3$ (40%). (c) PPTS, MeOH. (d) (i) TMSOTf, DIPEA; (ii) ZnBr, H_2O (72% from **660**). (e) (i) acetone, $TiCl_4$, TMSOTf; (ii) Ac_2O, DMAP (77%). (f) (i) $N_2H_2.H_2O$; (ii) $Pb(OAc)_4$. (g): (i) *hν*; (ii) HF•Pyridine (78% from **662**). DIPEA: diisopropylethylamine

hemiacetal bridging C-9 and C-13, for which conditions not affecting a similar protection of the 20-hydroxy group were developed (pyridinium-*p*-toluenesulfonate (PPTS) in methanol), thus obtaining the dimethylacetal **661**. As a prelude to the introduction of the three missing carbons, the tertiary hydroxy groups at C-3 and C-9 were protected as TMS ethers, and the C-13 ketone **662** was obtained by being chemoselectively deacetalyted with the Lewis acid zinc bromide. The stage was now set for the addition of the three missing carbons by $TiCl_4$/*n*-Bu_3N/TMSOTf-mediated cross-aldolization with acetone. The reaction occurred with remarkable selectivity on the methylene at C-14, when compared to that at C-12, and, after crotonization with Ac_2O-DMAP, the rhamnopholane enone **663** eventually was obtained. The selectivity of the enolization could be due to the major kinetic acidity of the C-14 *axial* proton, due to 1,3-diaxial interaction with the 9-oxygen. To complete the synthesis, the final steps of Wender's semi-synthesis of prostratin (**192**) were used [295]. Thus, the α-isopropylidene ketone **663** was reacted with hydrazine, affording, in a Knorr fashion, a Δ^5-pyrazoline derivative, not isolated but directly oxidized with lead tetraacetate to the C-13 acetoxylated diazene **664**. Photolysis and deprotection eventually afforded (+)-prostratin (**192**) in an overall seventeen steps from 3-hydroxy-2*H*-pyran-2-one (**657**).

5.5 *Summary of the Synthesis Aspects*

The intense synthesis activity on tigliane diterpenoids has produced many entries into core versions of their carbon skeleton and functionality, of which four have been finalized to natural products (phorbol (**1**) and prostratin (**192**), two total syntheses each). These achievements have identified the major challenge facing the duplication of Nature, namely, the formation of the hydroxylated C/D ring junction, capitalizing on ingenious solutions for the establishment of the C-12 *β*-hydroxy group and the hydroxylation of rings A and B by allylic oxidation of olefin precursors. In semi-synthesis studies, Wender also described a general strategy to cyclize rhamnopholane precursors into 13-hydroxylated tiglianes, deftly exploited in the two published syntheses of prostratin (**192**).

The syntheses by Wender have placed the novelty bar very high in the realm of the total synthesis of phorbol (**1**). They are based on functionalized precursors, and suffer from lengthy but generally efficient functional group transformation steps. Conversely, the Baran synthesis relies on C-H functionalization reactions, and, by their careful timing, functional group manipulation managed to minimize the step-count, at the expense, however, of overall yield, since oxygenation reactions have generally modest yields. The improvement in efficiency in the Baran synthesis (from 52 to 19 steps) has taken advantage in its final steps from findings associated with the Wender syntheses (establishment of ring B functionalization and stereocontrol in the formation of the 12*β*-hydroxy group). A novel synthesis of phorbol (**1**) unrelated to the Baran synthesis in regard to formation of the C/D-rings and to the Wender synthesis for the functionalization of the A/B rings will therefore require considerable ingenuity and command of organic reactivity.

Remarkably, the biogenetic Jakupovic hypotheses for the coupled formation of both the B/C and the C/D ring junctions by a cyclopropanone rearrangement has not yet been investigated for its potential to streamline the synthesis of the tigliane skeleton. The Wender and the Baran synthesis differ, apart from the way the pre-tigliane construct is built, also in the establishment of the critical hydroxycyclopropane moiety. Wender capitalizes on a cheletropic carbene addition to an acetoxyenone precursor, while Baran relies on the nucleophilic displacement of C-15 trifluoroacetate by a C-12 enolate.

Overall, the extensive knowledge on the build-up of proto- or pretigliane assemblies has so far been unexploited in terms of biological translation. The availability of high-resolution X-ray crystallographic data on the binding of phorbol esters to protein kinase C (see Sect. 6.1) could revive this area, combining expertise in synthesis and modeling to explore the protein kinase C pharmacophore and taming it to medicinal applications. Recent work by Williams on D-inverted phorbol esters exemplifies this fertile trend [336], which could benefit from the application of novel synthetic techniques (photocatalysis, electrosynthesis, etc.) to expedite the access of biologically relevant tigliane frameworks.

6 Biological Activity

Tigliane esters might well be the family of plant natural products boasting the largest number of citations in the biomedical literature, but their clinical gold remains still largely unmined. Several reasons underlie this situation. To start with, most bioactivity studies on tigliane derivatives have focused on a single type-A phorbol ester (phorbol myristate acetate = 12-tetradecanoylphorbol-20-acetate, (**2**)), for which data at the cellular-, animal-, and clinical levels are all available, and for which the biological profile is assumed to be mainly recapped by sub-nanomolar interaction with a specific series of isoforms of a single molecular target, protein kinase C (PKC) [7]. Even within this reductionist scenario, modifications of the diterpenoid core and/or its acyl decoration can modulate the spectrum of PKC isoform interaction, with a significant deviation from the biological profile of phorbol (**1**), as exemplified by the kinomic profile of the epoxytigliane tigilanol tiglate (**3**), a type C-diester oxidatively modified in ring B [8]. In addition, PKC-related [337] as well as PKC-unrelated non-kinase targets [338] have also been identified for phorbol esters, and novel ones might well emerge in the near future, providing an explanation for the often incomplete correlation between data from PKC-based assays and in vivo experiments.

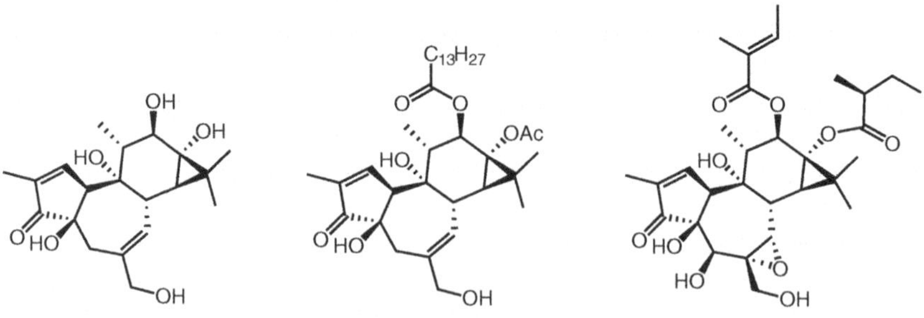

1 (phorbol) **2** (phorbol myristate acetate) **3** (tigilanol tiglate)

Next, many studies on the bioactivity of phorbol myristate acetate (**2**) were not done with the purified compound, but with croton oil, which, in addition to **2**, also contains tigliane esters of the B- and C-type as well as esters based on 4-deoxyphorbol [42]. An even more confusing case exists with the toxicity studies on *Jatropha* kernel meal, a protein-rich matrix that contains esters of polyols different from phorbol as well as toxic constituents of a different nature [339]. Due to the inavailability of reference tigliane standards, these studies have assumed that **2** is responsible for the poisonous effects of the plant biomass, of which the overall toxicity is currently expressed in terms of **2**-equivalents [339]. This simplification has practical relevance in a toxicological context aimed at establishing limits of exposure, but the interpretation of these studies outside a regulatory toxicological framework requires caution. As a matter of fact, it has been suggested that, rather than by tigliane esters, the highly poisonous nature of *Jatropha* seed cake should be associated with long-chain hydroxyacids isomeric with ricinoleic acid. These compounds down-regulate the uncoupling protein UCP3 and disrupt mitochondrial function [340], and tigliane esters could only be responsible for certain phenotypic aspects of the spent seeds (e.g. skin-irritancy, unpalatability).

The literature on the bioactivity of tigliane esters is large, and PMA (**2**) is a serious candidate for the most cited plant natural product in the whole biomedical literature, the veritable equivalent of what DNA is for biological macromolecules. In addition, the study of tigliane esters spans areas of both basic and applied relevance, ranging from medicine to esthetic surgery, and from environmental and industrial toxicology to nutrition and the green economy. To distill these scattered biological studies into coherent subsections, they have been classified according to their cellular-, animal-, or human nature.

6.1 Molecular Studies

The best investigated proteins selectively targeted by tigliane esters are a series of isoforms of the serine/threonine kinase protein kinase C (PKC) and of the transient receptor potential vanilloid channels (TRPV-1 and TRPV-4). Distinct structure-activity relationships have been demonstrated for these three targets, supporting the view of an orthogonal relationship between the kinase and the ion channel endpoints. The modulation of a host of additional macromolecular targets involved in inflammation and the immune response (NF-κB, MAP kinases, COX enzymes) is secondary to the phosphorylating activity of PKC and will not be discussed further herein.

Protein Kinase C

The serine-threonine kinase "protein kinase C" (PKC) was discovered in 1979 at Kobe University by Nishizuka in bovine brain [341]. Three years later, in studies on platelets, he and his French collaborator Castagna identified phorbol myristate acetate (**2**) as an ultra-potent biological analogue of unsaturated *sn*-1,2-diacylglycerols (DAG for short), the natural activators of this enzyme [342]. Protein kinase C and phorbol esters have then become closely connected, and, since animal studies had initially shown a correlation between PKC activation and positivity in the Berenblum tumor promotion assay (see Sect. 6.2), cancer has also been associated to both this enzyme and its exogenous agonists [7]. Seven families of effector proteins hosting a DAG-binding domain (the C1 domain of PKC) were next discovered, including the chimaerins (Rac GTPase activating proteins), RasGRPs (exchange factors for Ras/Rap1), and Munc13 isoforms (scaffolding proteins involved in exocytosis and synaptic transmission) [337]. The count of human proteins containing a C1 domain has now passed 66 [343]. Just like PKC, these proteins bind phorbol esters with great affinity, and undergo subcellular redistribution (translocation) after binding, and interact with a diversity of effector and regulatory molecules [344, 345]. Interest in these non-PKC phorbol-targets peaked at the end of the twentieth century, but then somewhat declined compared to that on PKC [344], while the nature of the relationship between PKC and cancer was questioned, suggesting the possibility of a cancer-protective rather than a cancer-promoting activity [346]. After briefly reviewing the biochemistry of PKC, the molecular details of the interaction of PKC with tigliane esters will be summarized in the light of structure-activity relationships and crystallograpic studies, highlighting next the cancer-promoting vs. cancer-protective role of PKC and tigliane esters.

Protein Kinase C: Basic Biochemistry

Protein kinase C is a multigene family of lipid-regulated serine/threonine kinases that occur virtually in every cell, and act as a central hub of cellular pathways [343, 347]. Protein kinase C family members regulate general transducing pathways for cellular proliferation, differentiation, and metabolism, and also exert tissue-specific functions. Individual isoforms have specific phosphorylation targets, and show cell- and/or tissue-specific expression, and exhibit specific, sometimes even opposite, physiological functions [342]. All PKCs share a similar catalytic core, inhibited by intramolecular binding to a regulatory pseudo-substrate module that acts as a veritable molecular safety catch, rendering the enzyme inactive. Protein kinase C activation is, at a molecular level, basically the result of a conformational change that releases the pseudosubstrate from the catalytic site, removing auto-inhibition [342, 347]. Control of this process by endogenous ligands is the basis for the partition of the nine human PKC members into three subfamilies [346]. Classical ((c)PKC) isoforms (α, βI, βII, and γ) require calcium (II), DAG, and phosphatidylserine (PS) for activation. Novel ((n)PKC) isoforms (δ-, ε-, η-, and θ) require only DAG for activation, while activation of the atypical isoforms ((a)PKCs) (Mζ-ι/λ isoforms) is independent from calcium and DAG, although remaining sensitive to phosphatidylserine [347]. The structural basis for this taxonomy is the presence of specific ligand sites in the kinase regulatory domain which binds to the catalytic site and acts as a pseudosubstrate. The regulatory domain hosts distinct binding sites, referred to as C1 and C2, for, respectively, DAG and calcium. In members of the classical family, both binding sites are functional, and their occupancy leads to dissociation of the pseudo-substrate from the catalytic site, which is now free to exert its phosphorylating activity [342, 347]. In the regulatory domain of PKCs from the novel family, only the binding site for DAG is functional, while neither C1 nor C2 is functional in isoforms from the atypical family. Having a functional C1 domain, PKCs from the classic and the novel families are sensitive to phorboid esters of the tigliane, ingenane and daphnane structural types. These ligands bind C1 in lieu of DAG and trigger the conformational change that liberates the catalytic site from its pseudosubstrate [347].

After activation, PKCs are translocated by specific proteins, RACK, receptor for activated protein kinase C proteins to the plasma membrane, where their activity is sustained transiently by the phospholipase-dependent generation of DAG from phosphatidylinositol, or, in a persistent way, by the presence of tigliane esters [342]. Translocation to the nucleus also takes place and could precede or follow that to the cellular membrane, but the role of the kinetics involved is still largely unclear. Surprisingly, microgravity interferes with the translocation of activated PKC (even PMA (**2**)-bound) to the plasma membrane, and a decreased translocation of PKC-δ in lymphocytes has been associated with the immunodeficiency of astronauts, providing a cosmic trait to research on phorbol (**1**) [348].

Within the various isoforms, some seem more involved in specific responses to tigliane esters than others, especially for the immune activity. Thus, PKC-β is essential for development and maturation of B-lymphocytes and for their production of immunoglobulins [349], and the mitogenic activity of PMA (**2**) on B-lymphocytes is used in the diagnosis of leukemias (see Sect. 6.4). The non-specific stimulation and activation of immune cells by **2** is mediated by the MAPK kinases cascade [349], and could be critically involved in the wound-healing and tissue regenerative activity by croton oil peel (see Sect. 6.4). In addition, the novel kinase member PKC-θ is the only member of the PKC family associated with the immunological synapse between T cells and antigen-presenting cells [349]. The functional interaction between PKC-θ and various downstream effectors mediates T-cell activation, differentiation, and migration. In addition, PKC-θ has also a nuclear activity of immunological relevance, inducing gene expression essential in the development of memory CD4+ T cells, up-regulating HIV-1 transcription, and raising the prospect of eradication of HIV-1 from its reservoirs by the combined use of PKC-θ activators and anti-HIV drugs (see Sect. 6.3) [349]. Within the non-immunological activities associated with PKCs, one of toxicological relevance is the regulation of the transcription of the insulin gene, for which its dysregulation by phorbol esters markedly affects blood glucose levels [350].

Other activities of PKCs seem less specific and general within the various isoforms. One is their susceptibility to activate by oxidative stress, presumably via the generation of DAG in the phospholipase C (PLC)-dependent cleavage of phospholipid hydroperoxides, and to play a critical role in the development of the clinical signs of inflammation [346]. It is not surprising therefore that the systemic effects of tigliane esters after the accidental consumption of one of their plant sources includes irritation of the mucosa of the intestinal tract, hemorrhaging, and necrotic organ lesions (see Sect. 6.4). Some of the most severe and non-inflammation-related effects of the systemic activity of tigliane esters could be related to modulation of insulin secretion [350] and to central nervous system activity, since DAG is involved in the signal transduction of the adrenergic, μ-cholinergic, and glutamate receptors [351].

The Interaction of Tigliane Esters with Protein Kinase C

There are two basic approaches to investigate the binding mode of small molecules to macromolecular targets. The first is the study of structure-activity relationships, which, with the aid of modeling, can lead to the definition of a complementary site on the macromolecule. The second approach is the direct study of the interaction by crystallographic methods or cryo-electron microscopy. However, neither is devoid of limitations. Structure-activity relationships are not "absolute", that is, different assays can produce read-outs that are not overlapping, while macromolecular targets are not frozen conformationally in solution, being capable of attaining different conformations and able to bind a ligand in more than one way.

The debate on the binding mode of tigliane esters to PKC exemplifies these subtleties. An incomplete overlapping in structure-activity relationships exists for tigliane esters between biochemical experiments (displacement of a radioligand from PKC, phosphorylation of substrates, membrane translocation) and phenotypic assays. In addition, discrepancies can exist between the read-outs of different phenotypic assays (tumor promotion, skin irritancy, HIV-1 latency reversal) or, as regards modeling, from the different database of biological analogues used. Some of these discrepancies are presumably due to in vivo esterase activity, which selectively removes a 20-ester group and unveils the activity of cryptic tigliane esters, but others have a mechanistic basis, like that between tumor promotion and skin irritancy (see Sect. 6.3), or else might be related to the profile of membrane translocation of activated PKC, with a different rate of transfer between the plasma membrane and nuclear translocation [352]. Most confusing, PKC activation has often been associated by default with tumor promotion, that is, to positivity in the Berenblum assay, despite an imperfect correlation between the two activities.

The first structure-activity relationships of phorbol esters were established in the realm of tumor promotion [342], while the first X-ray crystallographic data obtained were those of the complex between the PKC-δ C1 domain and phorbol 13-acetate (**38**)) [353]. This study identified a complementary image of this ester on the macromolecular receptor, but the structural details of this interaction turned out not to overlap with those derived from the structure-activity studies in the realm of tumor promotion. On the one hand, activity in tumor promotion assays is incompletely associated with PKC interaction, but concerns were also raised as to the biological relevance of the X-ray data, since the ligand phorbol 13-acetate (**38**) is an extremely weak activator of PKC [353]. Supported also by computational studies, distinct binding models have emerged from the pharmacological and the crystallographic studies on the interaction of phorbol esters and PKC. Both models acknowledge that ester decoration is essential for the activity, and that specific oxygen functions are required for binding to the C1 motif of PKCs through a network of complementary hydrogen bonding interactions. These eventually turn a hydrophilic surface on the C1 site into a hydrophobic one and trigger the conformational reorganization that eventually removes C1 from the catalytic site of PKC. However, the two models differ in terms of the identification of the oxygen functions involved in the network associated with this polarity swap, the Rubicon being the role of the hydroxy groups at C-4 and at C-9. The Hecker-Rando-Kishi (HRK) model of PKC activation is based mostly on phenotypic data of tumor-promotion across different classes of promoters, and associates activity with the combined presence of three oxygen atoms separated by approximately 6 Å (the C-9 and C-20 hydroxy and the C-3 carbonyl groups), and of a lipophilic acyl moiety bound at the C-13 hydroxy group [354, 355]. The Blumberg model is based on the X-ray analysis of phorbol-13-acetate (**38**) bound to the C1 domain of PKC-δ [354], and in this model the role assigned to the 9-hydroxy group in the HRK model is played by the 4-hydroxy, with the 9-hydroxy group not being directly involved in binding [356]. In addition, the contribution of the hydrophobic interactions in which the acyl moiety is engaged remains undefined.

A recent X-ray crystallographic analysis of the potent activator phorbol 12,13-dibutyrate (**503**) and the less potent activator prostratin (**192**) bound to the C1 domain of PKC-δ in the presence of the phospholipid 1,2-diheptanoyl-*sn*-glycero-3-phosphocholine, confirmed the Blumberg model in terms of location of the H-bonding network [357]. Thus, the 20-hydroxy group of the tigliane diester acts both as a H-bonding donor and two-fold acceptor, interacting overall with three aminoacyl residues of the PKC DAG-binding site, and the 4-hydroxy roup acts as a H-bonding donor, and the oxygen carbonyl at C-3 as a H-bonding acceptor. The 9-hydroxy group is not directly involved in binding, but, by forming an intramolecular H-bond with the carbonyl group of the acyl moiety at C-13, it behaves as a type of straightjacket for the conformational mobility of the C-13 ester group, helping in the attainment of an optimal geometry to maximize hydrophobic interactions [357]. Also the lipophilic element of the 12-ester group is directly involved in binding, but its role seems less critical, since deoxygenation at C-12 does not decrease affinity critically [357]. A grapical impression of the X-ray structural data of **503** and **192** bound to the C1 domain of PKC-δ together with two and one molecule of 1,2-diheptanoyl-*sn*-glycero-3-phosphocholine is provided in Plate 8.

Comparison of the X-ray crystallographic data on the complex of phorbol 13-acetate (**38**) and phorbol 12,13-dibutyrate (**503**) with PKC highlights the relevance of hydrophobic interactions in small molecule-macromolecule interaction. Both compounds bind PKC with the same network of H-bonding interaction, but **38** is almost unable to activate PKC, while **192** is a one-digit nanomolar activator [357]. The relevance of "lipophilic" aliphatic carbon-hydrogen bonds for non-covalent ligand-protein interaction is exemplified dramatically by biotin (**665**), for which the femtomolar affinity for avidin was shown by X-ray crystallographic analysis to be associated to eighteen intermolecular interactions. Six of these involving the aliphatic C-H bonds of the valeryl moiety of biotin, which form CH–π H-bonding with the aromatic aminoacidic residues of avidin, significantly contributing to binding [358].

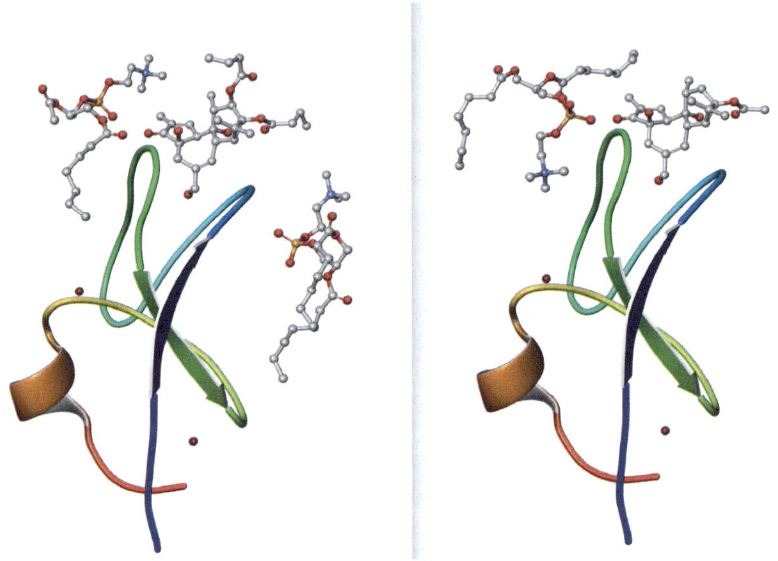

Plate 8 Three-dimensional model of **503** (left) and **192** (right) bound to the C1 domain of PKC-δ together with two, and one molecule, respectively of 1,2-diheptanoyl-*sn*-glycero-3-phosphocholine. The structures were derived from the data in Ref. [357] deposited in the Protein Data Bank in Europe (https://www.ebi.ac.uk/pdbe/) via accession numbers 7KNJ and 7LCB

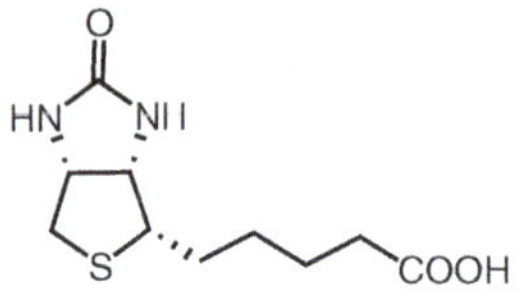

665

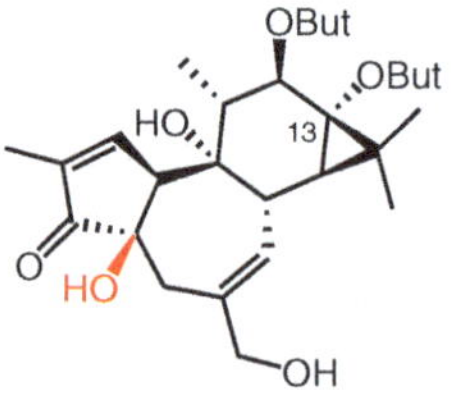

503

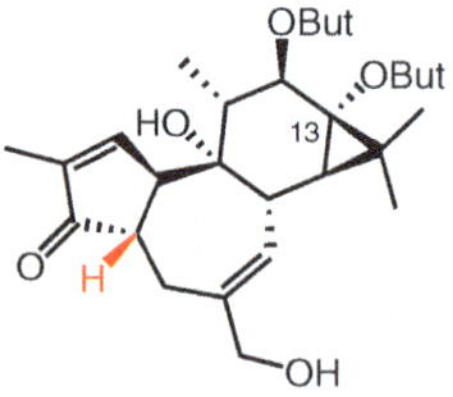

666

The contribution of the 4-hydroxy group to the binding of phorbol esters to PKC remains unclear, since its removal in phorbol 12,13-dibutyrate (**503**) to its 4βH-deoxy derivative **666** had a negligible effect on PKC affinity, as measured by radioligand displacement, as well as by the generation of superoxide and the induction of the Epstein-Barr virus [359]. A similar observation was done in a study on platelet aggregation, where only a modest erosion of activity was associated with removal of the 4-hydroxy group [115]. The X-ray binding data also lack the emphasis on the intramolecular H-bonding between the carbonyl ester bound to the 13-hydroxy and the tertiary 9-hydroxy groups, a critical element of the HRK model also observed in REDOR NMR [360] and a solid-phase study on phorbol 12,13-diacetate (**542**) and the PKC-δ C1b domain [361]. The relevance of this intramolecular hydrogen bonding is highlighted by the detrimental effect on the PKC affinity associated with the presence of a free 12*β*-hydroxy group, which competitively perturbs the pattern of hydrogen-bonding engagement of the 13-ester carbonyl and, as a consequence, the orientation of the lipophilic element of the acyl moiety. The differences existing between the solid-state conformation of C1 in solution and in the solid state could be due not only to the inherent dynamic properties of proteins [362], but also to the lipid microenvironment where PKC translocates after binding to tigliane esters, blurring the translation of structure-activity relationships into binding model(s).

Despite these ambiguities, the single conformational state of the crystallographic analysis rationalizes very well with the other structure-activity relationships. Thus, methylation of the C-4 hydroxy group, in addition to rendering the 4-hydroxy as being unable to act as a H-bonding donor, also interferes by steric hindrance with the interaction of the 4-carbonyl, resulting in a complete loss of activity. Given the critical role of the 20-hydroxy group as H-bonding donor and acceptor, its alkylative or acylative modification causes loss of activity. While a free 12-hydroxy group decreases activity, interfering with the lipophilic interaction of the 13-acyl group, deoxygenation at C-12 is tolerated, with activity depending on the lipophilicity of the ester group at C-13. On the other hand, skeletal remodeling by opening of the cyclopropane ring as well as isomerization at C-4 upsets the delicate network of interactions involved in PKC binding, and is associated with loss of activity.

Epoxidation of the ring-B double bond is associated with loss of activity when the oxirane oxygen is *β*-oriented, possibly because of a combination of intramolecular H-bonding with the 4-hydroxy group, which displaces the 20-hydroxy from an optimal orientation for binding [45]. This diversion of H-bonding is reminiscent of that associated with the presence of a free 12-hydroxy group on the activity of the 13-esters of phorbol (**1**) [357]. Competition of H-bonding donation to the ester carbonyl between the hydroxy groups at C-9 and C-12 leads to a marked decrease of activity [359]. On the other hand, the combined presence of an *α*-epoxide and of a 5*β*-hydroxy group is tolerated, and associated with a remodulation of the profile of PKC activation [8, 45]. Thus, while PMA (**2**) activates all classic and novel kinases, in translocation assays tigilanol tiglate (**3**) behaves as a selective isozyme modulator, retaining activity with the classic isoform PKC-β, while having modest activity on PKC-α and PKC-γ, and marginal activity on other isoforms [8, 45]. In principle, the presence of a 5-hydroxy group potentially may have a critical effect on PKC binding, since it could engage the hydroxy groups at C-4 and C-20 in intramolecular H-bonding, undermining the network of H-bonding underlying PKC binding. However, comparison of TT (**3**) and its ring B olefin analogue (**666**) showed that the presence of the 5-hydroxy group alone recapitulated the selectivity of PKC observed for the natural product [45].

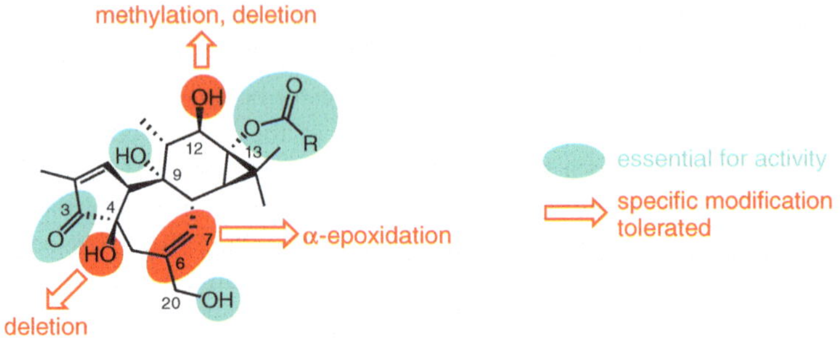

Fig. 10 Structure-activity relationship for the interaction of tigliane esters with protein kinase C

Figure 10 summarizes the current consensus on the structure-activity relationships for the interaction of tigliane esters with PKC, distinguishing between modulating and critical elements for activity. Remarkably, PKC activation does not necessarily translate into activity in specific phenotypic assays, and in particularly on that of tumor promotion.

The Cancer Connections of Protein Kinase C and Tigliane Esters

The advent of cancer genomics has shown that PKC is often mutated in cancer cells, especially in melanoma, colorectal cancer, and lung squamous cell carcinoma, where aberrant forms of the enzyme are found in almost one fourth of tumors [6, 346, 362]. This is hardly unexpected, given the critical role played by PKC in cell differentiation and proliferation as well as in tumor progression [363, 364], but, surprisingly, most of these mutations are associated with complete functional loss, in sharp contrast to those associated with Alzheimer's diseases, which are, conversely, mostly of the gain-of-function type [365]. Protein kinase C traditionally has been associated with positivity in the Berenblum skin tumor promotion assay (see Sect. 6.3), but doubts have been raised on this association, since positivity in the assay can be induced by unspecific tissue damage, such as when caused by exposure to ultraviolet light, radiation, chemical irritation or mechanical wounding. In addition, the association between tumor promotion and PKC activation is far from perfect, as potent PKC activators like the bryostatins [366] and ingenol mebutate (**9**) [367] are not tumor promoters, while also staurosporine, a PKC pan-inhibitor, shows tumor-promoting activity, and PKC inhibitors can only at best mitigate but not prevent tumor promotion [368]. Tumor promotion aside, it is, in any case, difficult to assign a cancer-protective or cancer-supportive role to PKC, since its oncogenic/oncosuppressor role is isoform- and tumor-type dependent [6].

Thus, while PKC-ι acts essentially as an oncogene in a variety of cancers, there is strong evidence that PKC-α is a tumor suppressor in colorectal- and lung cancers, and evidence that enhancement of PKC-α activity can be beneficial in many cancers has also been reported [6]. To add a complication to this scenario, the hyper-stimulation of PKC by PMA (**2**) and related tigliane esters leads to chronic down-regulation of this enzyme, associating a fast stimulatory activity with a longer term form of inhibition [6, 364]. An additional source of uncertainty stems from the use of murine cancer models, since modulation of PKC by phorbol esters in various strains of these rodents is associated with tumor promotion, a mechanistically unclear concept which, although not relevant for humans at least in terms of the Berenblum assay (see Sect. 6.3), suggests nevertheless the existence of unique connections between PKC and cancer in these animals [6, 346].

Taken together, these observation show a complex scenario, where the "good" or "bad" role of PKC on cancer cells is isoform- and tissue dependent, and that of tigliane esters on PKC is biphasic, with a fast activation followed by a longer desensitization. This is in line with the observation that PKC-based anticancer compound screening campaigns based on an inhibition strategy failed to demonstrated clinical efficacy, even worsening in some cases the outcome of the disease [346]. These efforts were based on ATP competitive inhibitors (staurosporine and analogues) to deactivate the catalytic site of PKC, or on antisense-oligonucleotides to reduce the expression levels of PKC isoforms. Clinically beneficial effects were only observed with midostaurin (**7**), but were related to the tyrosine kinase inhibitory activity of this compound [346, 365]. The macrolide bryostatin shows sub-nanomolar affinity for the C1 regulatory site of PKC targeted by DAG and phorbol esters, promoting activation of inhibition depending on short- or long-term exposure [366, 367], but also this compound failed to demonstrate clinical benefits either in monotherapy or when combined with other anticancer drugs [365].

Despite these setbacks with non-tigliane PKC modulators, beneficial clinical results have been obtained with phorboids. In 2012 a gel formulation of the ingenane ester ingenol-3-angelate (Picato®) (**9**), a powerful PKC agonist, was approved as a topical chemotherapeutic agent for the treatment of actinic keratosis by both the EMA in Europe and the FDA in the U.S. [34]. However, after reviewing the literature on the safety of the product, in 2020 the the EMA concluded that when compared to the immunosuppressor imiquimod, ingenol mebutate increased the risk of skin cancer, and the product is no longer available in Europe. In 2020, an intratumoral injection of the tigliane ester tigilanol tiglate (Stelfonta®) (**3**) [8] was approved by the U.S. FDA and by the EMA for the treatment of non-metastatic mast cell tumors in dogs, and currently the compound is undergoing clinical development for the management of melanoma and head-and-neck cancer (see Sect. 6.4). The precise mechanisms of action of ingenol mebutate (**9**) and tigilanol tiglate (**3**) are not known in detail, but both compounds show a more selective spectrum of PKC isoform activation than the PKC pan-activator PMA (**2**) [8, 34]. The cytotoxic activity of tigilanol tiglate (**3**) has been shown to be PKC-dependent [8], and its clinical efficacy has been suggested to result from a combination of tumor blood vessel destruction and direct killing of tumor cells, with a strong involvement of the immune response.

The uncertain nature of the relationship between PKC modulation and cancer (tumor promoter or tumor suppressor?) parallels that between free radicals and this disease type. Following a seminal article by James Watson of DNA fame [368], it has been suggested, against current orthodoxy, that free radicals and reactive oxygen species could have a protective role in cancer, contributing to correct protein folding in the endoplasmic reticulum and the maintainance of their functionality [369]. Conversely, cancer cells have a shortage of reactive oxygen species, due to the Warburg effect that de-emphasizes the Krebs cycle, a major source of reactive oxygen species (ROS), at the expense of glycolysis, and, conversely, generates the powerful antioxidant NADH. According to this theory, the beneficial effects of exercise on cancer could be due to the generation of free radicals in contrast to the shortage of ROS due to the metabolic rewiring and deregulated energetics of cancer cells. The generation of reactive oxygen species is also the first-line of defense against toxic compounds, and regular injections of vitamin E, a powerful antioxidant, have been reported to increase, and not to decrease, the incidence of lung and prostate cancers in humans [368, 369]. Tigliane esters, major generators of ROS via the activation of PKC, could then be viewed as the cellular equivalent of physical exercise.

Just like ROS, also PKC and tigliane esters exemplify how many aspects of cancer, if not terra incognita, still remain dubious and provide a reminder that chemicals per se are not good or bad, but become so according to their use. It does not seem inconceivable, therefore, that also the effects of PKC activation by tigliane esters depend on the cellular context and the time in which they take place. In the context of cancer, these compounds are therefore neither good nor bad, but both.

Transient Receptor Potential Ion Channels

Owing to the presence of a cysteine-rich ankyrin moiety sensitive to activated oxygen species, all transient receptor potential ion channels are subjected to activation by tigliane esters in an indirect way, that is, via PKC-dependent generation of free radicals. This activation could be of great relevance for the wound-healing activity of tigliane esters, since TRPV1, TRPV3, and TRPV4 are expressed in skin keratinocytes and have been shown to stimulate wound healing in a variety of cellular and animal models (see Sect. 6.4) [370–372]. In addition, tigliane esters can directly bind and activate two specific types of thermo-TRPs, TRPV4, and TRPV1.

The Ion Channel Protein TRPV4

The ion channel protein TRPV4 is a calcium-permeable, nonselective cation channel broadly expressed and involved in a range of organ-specific diverse activities, which include the brain regulation of osmotic pressure, the skin response to UV-B light, as well as liver-, intestinal-, renal-, joint-, lung-, and bladder functioning [373]. The ion channel protein TRPV4 is sensitive to both physical (osmotic, mechanical, non-noxious heat) and chemical activation [374], and epoxyeicosatrienoic acid from the oxidative metabolism of arachidonic acid and anandamide has been identified as a possible endogenous activator [375]. Despite broad expression, a diversity of activities, and polymodal activation, TRPV4 has been given considerable attention from a medicinal chemistry standpoint. Thus, it plays a critical role in chondrocytes, the cells responsible for extracellular matrix (ECM) turnover, for which their mechanotransduction by TRPV4, Piezo1, and Piezo2 influences cartilage maintenance, degeneration, and regeneration [376], and the local injection of TRPV4 modulators has received considerable attention for the management of osteoarthritis [376]. There is also growing evidence for a druggable involvement of TRPV4 in inflammation [373], pain [377], and wound-healing [378]. Of recent topical interest, the SARS-CoV-2 spike protein induces pulmonary vascular endotheliopathy in a TRPV4-mediated fashion [379].

4α-Phorbol diesters have emerged as remarkably selective activators of TRPV4 [380], with potency peaking at the level of 12,13-dihexanoate (**667**) and the 12,13-didecanoate (**668**) [380]. 4α-Phorbol diesters show marginal, if any, activity in PKC activation assays, and PMA (**2**) was at least 50-fold less potent than 4α-phorbol 12,13-didecanoate (**668**) for the activation of TRPV4, a strong indication that the affinity for PKC and TRPV1 are related orthogonally to the tigliane scaffold. Experiments using site-directed mutagenesis, calcium imaging, and electrophysiology localized the binding site of 4α-phorbol diesters in a domain loop of TRPV4, analogous to the binding of capsaicin in TRPV1 [380, 381]. By analogy with binding to PKC, the lipophilic ester groups on ring C critically affects the orientation of the tigliane core into the ligand binding pocket. However, the relationship between side-chain length and activity was not linear, with activity greatest at both a C-6 and a C-10 carbon length, in sharp contrast with the bell-type curve of potency vs. acyl moiety length observed for the activation of PKC by phorbol diesters [380]. The bisdecanoate **668** was used to characterize TRPV4-expressing cells in the bladder urothelium, suggesting that this ion channel could be a target for the treatment of urothelial disfunction [382]. Also, *lumi*-4α-phorbol didecanoate (**669**) can activate TRPV-4, albeit less potently than its corresponding ester from the 4α-phorbol series (**668**), but patch-clamp experiments showed a distinct kinetic of current decay, hinting at the possibility of prolonged activation of the ion channel, an observation of significant relevance for bladder dysfunction. Removal of the 12-ester group decreased the activity of 4α-phorbol 12,13-didecanoate (**668**), and a complete loss of activity was observed for its corresponding parent polyol (lumiphorbol, **500**) [380].

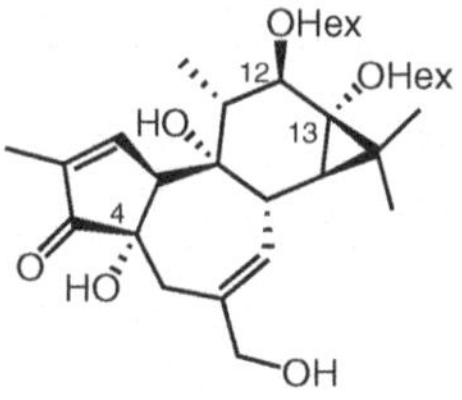

667

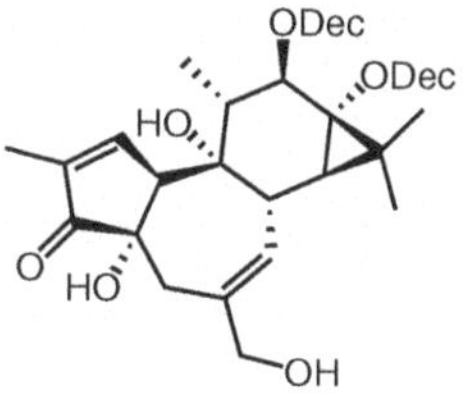

668

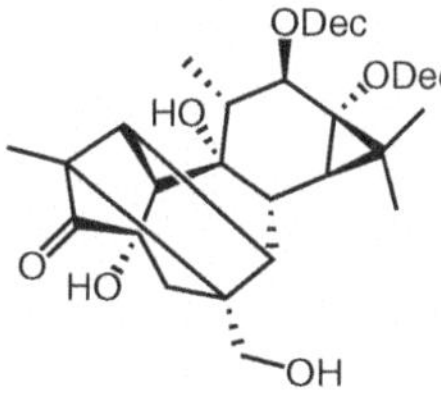

669

Transient Receptor Potential Vanilloid 1

Transient Receptor Potential Vanilloid 1 (TRPV1), better known as the capsaicin receptor, is the most well-investigated member of this family of ion channels [383]. The daphnane phorboid resiniferatoxin (RTX) (**10**) is an ultrapotent biological analogue of capsaicin, and, by intra-articular administration, currently is undergoing clinical development for the management of knee pain [384]. Owing to the acylation of the 20-hydroxy group, RTX does not bind to PKC and, despite its ultrapotency in the mouse ear erythema assay, it is not a tumor-promoter [384]. Owing to the difficult accessibility of RTX, an attempt was made to use phorbol as a scaffold to induce TRPV1 activity by decorating it with the acyl moieties of RTX. These efforts eventually produced the tigliane derivative, phorbol-12-phenylacetate-13-acetate-20-homovanillate (**670**) [385], with one-digit micromolar affinity for TRPV1 and a powerful desensitization activity on this ion channel [386]. However, the binding mode of **670** to TRPV-1 is different from that of RTX and capsaicin in terms of cooperative behavior as well as species selectivity, since **670** can bind to rat TRPV-1 more strongly than human TRPV-1, a difference that was traced to a single mutation (methionine to leucine) in the aminoacidic sequence of the two ion channels. The replacement of a linear amino acid (methionine) with a branched one in hTRPV1 reduced the ability of **670** to activate the ion channel by approximately 20-fold [387]. Owing to these properties, **670** proved useful in studies on the comparative pharmacology of this ion channel and the manner in which it is activated by small molecules.

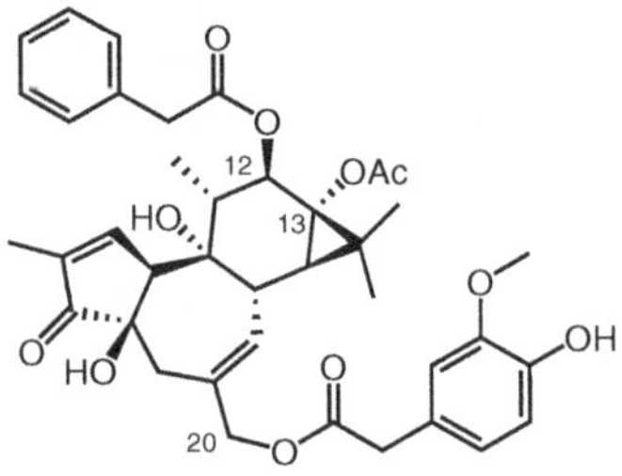

670

The remarkable biological profile of the 4α-phorbol diesters **667**, **668**, and of the 20-acyl phorbol **670** to selectively target TRPV-type ion channels suggest that, despite a significantly distinctive potency range (μM vs. nM), the tigliane motif nevertheless has the potential to go beyond activities related to the activation of PKC, and would be well worth a systematic investigation.

6.2 *Cellular Studies*

Animal Cells

Phorbol 12-myristate-13-acetate (**2**) is used extensively to investigate the cellular events associated with diacylglycerol (**671**), the endogenous activator of classic PKC isoforms. The broad range of biological activities makes **2** an invaluable biochemical tool for cellular studies, highlighting the clinical potential of PKC activators. Due to the universal role of PKC, interaction with different types of cells and induction of distinct activities has been demonstrated for tigliane esters. Given the vast nature of the literature, only activities of relevance for potential translation into animal and/or clinical studies, will be discussed.

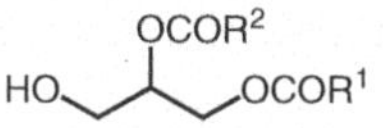

671

$R^1 = C_nH_{2n+2}$

$R^2 = C_nH_{(2n-2x)^*+2}$

* x= 0-2

Cancer-Related Cellular Studies

A major use of phorbol 12-myristate-13-acetate (**2**) is the PKC-induced induction of endogenous superoxide production by the activation of NAPH-oxidase [388]. The result is a declining mitochondrial membrane potential and a decreased activity of the respiratory chain complex IV, overall resulting in nanomolar cytotoxicity for several cell lines [389]. Owing to diversity in the cellular expression of PKC, it is difficult to compare the cytotoxicity data on tigliane esters, since IC_{50} values can range from nano-to micromolar values depending on the cell line. Perhaps not surprisingly, a poor correlation between cytotoxicity data and in vivo activity has been observed [337, 345].

In cancer diagnostics, **2** is used in cytogenetic tests as a B-cell specific mitogen, and **2**-induced B-cell division is important for the diagnosis of cancers and especially chronic lymphocytic leukemia [390]. Thus, chromosomes are visualized in dividing cells, and **2** is used to stimulate B-cell division and differentiation in the diagnosis of this type of cancer. In combination with the intracellular calcium booster ionomycin, **2** is also used to stimulate T-cell activation, proliferation, and cytokine production [391]. Leukemia cells show impaired cellular differentiation, and induction of differentiation stops uncontrolled cell replication [391]. This, along with the stimulatory activity on immune response, has provided a cellular basis to investigate the clinical translation of the potential anticancer activity of TPA and of tigliane PKC activators in general.

Antiviral Drug Discovery

A second line of cellular activity-based clinical interest for tigliane esters came with the plague of HIV. Human immune deficiency spurred unprecedented research activity on antiviral agents, eventually resulting in the discovery of potent anti-retroviral drugs that made it possible to treat this infection. On the other hand, antiretroviral therapy is not curative, since latent-proviruses integrated into the genome of CD4+ cells remain capable of replication, and some of these cells can revert to a quiescent state, persisting for the whole life of the host as resting memory cells [392]. The infection is therefore only latent, since activation of these proviruses can restart the disease. This genomic reservoir is antiviral-therapy resistant, and the eradication of HIV in infected patients would therefore require its clearance [392]. From these considerations, a "shock and kill" strategy was conceived, based on the administration in HIV-infected patients of a delatentizing agent in combination with anti-retroviral treatment [392]. Immunological T cells activation proved too toxic [392], and PKC agonists were the only agents to demonstrate reproducible results across different latency models of latency [392]. Protein kinase C-mediated T-cell activation is based on an interaction with the transcription factor NF-κB. Protein kinase C inactivates this by phosphorylation of IκB, releasing its p65 subunit RelA. Transcription factor NF-κB can then enter the nucleus, bind to complementary sites in the viral reservoir, and induce its expression [393]. The PKC activators phorbol esters and bryostatins give, in principle, a one-two punch to the virus, not only activating HIV expression in latently infected CD4+ T-cells, but also preventing uninfected cells from becoming infected by down-regulating the expression of the HIV-1 cell surface viral co-receptors CXCR4 and CCR5 [394, 395]. Within the phorboids, interest has focused on PMA (**2**), 12-deoxyphorbol 13-acetate (prostratin) (**192**) and 12-deoxyphorbol 13-phenylacetate (**201**). Unlike **2**, these two 12-deoxyphorbol esters are inactive in assays of tumor-promotion and (potential and questionable) oncogenicity [142]. In addition, prostratin (**192**) was isolated from a euphorbiaceous plant (*Homalanthus nutans* (G. Forst.) Guill.) used in traditional Samoan medicine for liver diseases without any apparent toxic effects [142]. Prostratin (**192**) was actually first isolated from a toxic thymeleaceous plant (*Pimelea prostrata* Willd.) [145], and its safety in humans has still to be demonstrated. In addition, when a series of 13-acyl phorbol derivatives were assayed for viral activation in in Jurkat T-cells containing latent HIV proviruses, a direct relationship was found between the length of the acyl moiety and activity, with potency peaking with the 13-stearoyl ester and with **3**, the 3-acetyl ester, showing only 10% of its potency [396]. A similar relationship between potency and lipophilicity was also observed in lactonized analogues of DAG [397]. Nevertheless, since delatentization strategies with antibodies had faced severe safety issues [394], concerns exist on the systemic inflammation triggered by the delatetization process, and the human exposure to prostratin in folk medicine apparently was sufficient to qualify this compound as a reference delatentizing agent [394].

The development of prostratin (**192**) from folk medicine to a potential drug vividly exemplifies the legal issues associated with the exploitation of biodiversity [398, 399].

In the early 1980s, the ethnobotanist Paul Cox from Brigham Young University in the United States first documented the use of the stem bark of *Homolanthus nutans* for the treatment of hepatitis by two distinct Samoan healers [400]. Cox, a Mormon and a botanist by education, had first come to Samoa in 1973 as part of a two-year after-graduation compulsory missionary service. He came from an academic background that was actively involved in the study and conservation of biodiversity, and, in 1984, Cox settled with his family in the isolated village of Falealupo in Samoa [399]. Four years later, by even mortgaging his own house in Utah, he managed to raise money to protect a large area of rainforest around Falealupo from logging, earning the trust of the village [400]. Cox was critical in formulating in 1989 a benefit-sharing agreement (Falealupo Covenant) between the two healers from which he had first learnt the use of *H. nutans* in folk medicine, the chiefs of their village, and the government of Samoa. In compliance with this agreement, Cox provided *Homalanthus* samples to the National Cancer Institute (NCI) to be assayed for anti-HIV activity [400]. After the discovery of the outstanding anti-HIV activity of the extract of the plant provided by Cox and the identification of prostratin (**3**) as its major active constituent [395], the village chefs and the Prime Minister of Samoa agreed to a patent filed by the NCI and Brigham Young University, Cox's employer, on the use of **3** as an anti-HIV agent. In the case of licensing of the compound from NCI for pharmaceutical development, the licensing contract would involve a fair and equitable return to the Samoan people, as stated in a benefit-sharing agreement negotiated by the NCI and the AIDS Research Alliance in 2001, as signed by the Prime Minister of Samoa as well as by the two village chiefs [399, 400]. As a result of these agreements, activities worth over $500,000, which included the building of three schools and a small hospital were carried out in Samoa. Cox gained a chief title in Samoa (Nafanua), and in 1997 received the Goldman Environmental Prize, along with a village chief, for efforts to preserve the Falealupo rain forest [399]. A similar agreement was signed between the Samoan Government and the University of California Berkeley to provide legal support to a project led by Jay Keasling to identify the gene sequence involved in the biosynthesis of **192** [399].

Things then started to become more complicated subsequently [399, 400]. In 2009, Cox had to face the accusation that the patent on prostratin was illegal, since the two Samoan healers were not included as co-inventors, but was eventually cleared from the accusation by a document signed by 87 village chiefs. The year before, Paul Wender had devised an elegant synthesis of prostratin from phorbol [295], and in 2010 the AIDS Research Alliance signed an additional licensing agreement with Stanford University, securing the rights to the Wender semi-synthesis of prostratin [399] This was highly important, since the concentration of **192** in *H. nutans* is very low, in the range of 2–50 mg/kg in the richest organ (stem), casting serious doubt on the extent of human exposure to **3** via folk medicine and on the possibility that this plant, even within a context of selective breeding, could have sustained the clinical development of prostratin [401].

Potent HIV latency reversal activity was also discovered in 4-deoxyphorbol 12-tigliate-13-isobutyrate (**158**), a non-promoting type-C tigliane ester isolated from *E.*

semiperfoliata Viv., of Sardo-Corsican endemism [64, 402]. This compound outperformed prostratin (**192**), synergized with the HDAC inhibitor vorinostat, and was equipotent ith bryostatin-1 in terms of anti-HIV-1 activity. Structure-activity relationship studies highlighted the role of the lipophilicity of the acyl moiety, in accordance with the observations made on phorbol diesters [402, 403]. A detailed study of its activity showed evidence that HIV-delatentization and down-regulation of HIV co-receptors are mechanistically distinct events, possibly involving different targets [403]. Significant activity against the chikungunya virus was also observed for deoxyphorbol esters [402], while sub-nanomolar anti-HIV activity was reported for phorbol 12-benzoate-13-octanoate (**69**) [99].

As of today, the phase I human clinical trial of prostratin (**192**) planned by the AIDS Research Alliance has not yet materialized, the genetics of its biosynthesis remain unknown, and the clinical translation of HIV latency activation by PKC activators received a blow from the failure of a clinical trial on bryostatin-1 [404], a biological analogue of prostratin in terms of PKC activation [405]. Well-tolerated dosages of bryostatin-1 failed to achieve systemic detectable concentrations in most patients, and, in any case, did not induce any latency reversal [405]. The beneficial clinical effects of **192** have not yet materialized, but it remains the reference compound to evaluate HIV-latency-reversing agents [406] and their combination [407], while its story exemplifies how local communities in one of the poorest countries in the world could benefit from their biodiversity, providing a legal framework for future folk medicine-inspired drug discovery. It also focused phytochemical attention on euphorbiaceous plants from Polynesia and Australia [408], setting the stage for the discovery of tigilanol tiglate (TT (**3**)) (see Sect. 6.3).

Platelet Activation

Human platelets selectively express several phorbol-ester sensitive protein kinase C isoforms (α and β within the classic types, δ and θ within the novel forms), and their activation induces rapid aggregation and release of granules packed with bioactive compounds (ATP, ADP, serotonin, and various peptides and proteins), eventually leading to the formation of thrombi. It is unclear how this in potent (one- or two digit nanomolar EC_{50}) and, in principle, detrimental thrombotic activity, could adversely affect the clinical profile of tigliane esters, but it can be easily measured turbidimetrically, providing the opportunity for extensive structure-activity relationships [115]. Alternatively, the liberation of ATP can be quantified conveniently with a luciferin/luciferase assay, providing a quick and precise proxy for the selective activation of the four PKC isoforms expressed in platelets [409]. Thus, a clear-cut correlation was found between the PKC-dependent phosphorylation of various signaling molecules (as measured by Western blot analysis) and the liberation of ATP measured with a luciferin/luciferase assay, supporting the critical role of PKC in platelet aggregation. Since platelets lack a nucleus, the pattern of plasma membrane/nuclear translocation of PKC can not be investigated with a platelet model.

Remarkably, a poor correlation was found with tumor-promoting activity and platelet activation assays, since also non-tumor promoters like prostratin showed potent activity in the platelet assays.

Bacterial and Fungal Cells

The interaction of tigliane esters with non-animal cells was focused mainly on the action of living cells on tigliane esters rather than on the perturbation of cell architecture and metabolism by these compounds. These studies were spurred on by the mass cultivation of euphorbiaceous seed plants for biodiesel production [410]. This generates a protein-rich pannel, which, after removal of tigliane esters, can be used as animal feed or as a fertilizer. The microbial and fungal degradation of these compounds therefore has important practical implications, and was investigated with the *Jatropha* oil pannel, a matrix rich (up to 1 g/kg) in 12-deoxy-16-hydroxytigliane esters (*Jatropha* factors) [410, 411]. Incubation of the oil pannel with soil for five weeks led to to the complete disappearance of tigliane esters, as evidenced by chromatographic analysis as well and after validation in selected bioassays. In addition, no transfer of tigliane esters was observed in plants grown in the Jatropha oil cake-supplemented soil [411]. A specific esterase (KM109) was characterized from the Gram-negative bacterium *Acinetobacter calcoaceticus*, but the enzyme was not further characterized [410].

Also several macro- and microfungal species could efficaciously detoxify the *Jatropha* tigliane esters by hydrolysis. Within basidiomycete fungi, the edible fungus *Pleurotus ostreatus* showed an excellent detoxifying capacity, retaining only traces of tigliane esters and also increased the nutritional value of the cake by decreasing its lignin content [410]. Various ascomycete fungi, including common *Penicillium*, *Aspergillus*, and *Rhizopus* species, were found also to metabolize the tigliane esters from *Jatropha* [410]. It is quite possible that fungi and bacteria could also modify the diterpenoid tigliane skeleton by additional, non-hydrolytic pathways, as shown for several diterpenoids, but little information exists on this topic, which seems well worthy of conducting systematic studies.

6.3 Animal Studies

Tigliane esters have shown activity in animal models en route to their potential clinical validation as anticancer and antiviral agents. These studies, carried out with standard pharmacological models, will not be discussed further, focusing rather on tigliane ester-based animal studies of general relevance for pharmacological research, or of relevance for toxicology and veterinary investigations.

The Berenblum Tumor Promotion Assay

The first carcinogenic compound was identified experimentally in 1915 by Yamagiwa and Ichikawa, who demonstrated that the repeated application of coal tar to the ears of rabbits led to the development of tumors [412]. Yamagiwa expressed his satisfaction for the discovery in a 17-(Japanese) word haiku (three-line poems) which translates as "Cancer is identified! Proudly I walk a few steps" [413]. One wonders if Yamagiwa had an idea of the number of steps still necessary to arrive at a full comprehension of the molecular basis of cancer and to its clinical control, two goals still elusive more than a century after his seminal paper. Coal tar is an umbrella name, being a complex mixture of aromatic compounds formed from condensing the vapors associated to the production of coke from coal. Coal tar served as a critical material to support the development of academic and industrial organic chemistry in the second half of the nineteenth century, a role it gradually handed over to oil from the turn of the twentieth century [414]. Following the seminal report by Yamagiwa and Ishikawa, various carcinogenic polycyclic aromatic compounds have been found in coal tar, for which its cancer-producing activity is due mainly to 7,12-dimethylbenz[*a*]anthracene (**672**), with its parent arene also having been isolated from coal tar [415].

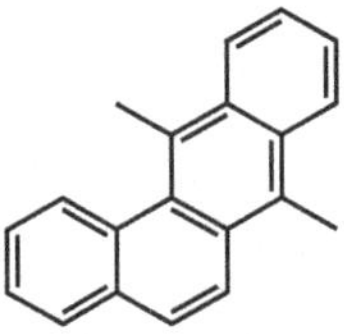

672

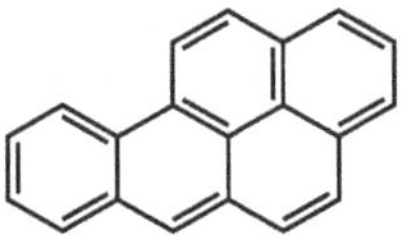

673

The mechanistic study of chemically induced carcinogenesis led to a two-stage theory for this process, and croton oil and **2** played a critical role in the development of this model of tumor development [20]. According to the two-stage theory of carcinogenesis, exposure to a carcinogenic compound damages DNA and induces the formation of latent tumor cells (initiation stage), which, after exposure to a hyperplasic agent (promotion stage), develop into a tumor [416]. The basic study on which the two-stage theory of carcinogenesis rests was reported in 1941 by Isaac Berenblum, who observed that the alternate application of croton oil and small dosages of benzo[*a*]pyrene (**673**) to the skin of mice induced a larger number of tumors than the same doses of benzo[*a*]pyrene alone [20]. Three years later, Mottram reported that subthreshold concentations of **673** to the backs of mice followed by a repeated weekly (biweekly) application of croton oil led, after 20–30 weeks, to the formation of malignant sarcomas [417]. The observations by Mottram were codified, using **672** rather than **673** as a cancer-inducing agent, in a tumor-promoting assay generally referred to as the Berenblum assay [418]. The sequence and time of application were both critical, since reversal of the order of application (first croton oil and then **672**) or extension of the interval of application of croton oil from weeks to months left the animal tumor-free. Croton oil, even after repeated treatments, could not induce the formation of sarcomas, but, occasionally it producednon-malignant benign papillomas [416]. Using this assay, other compounds were identified as tumor promoters, that is, those capable of promoting the development of malignancies after exposure to subthreshold dosages of a carcinogen, but their activity was two to three orders of magnitude lower than that of **2**, the "active" constituent of croton oil [419]. Among the tigliane esters, the mono- and diacyl derivatives of 4α- and 20-deoxyphorbol do not show promoting activity, nor do the 20-esters of 12,13-diacylphorbol and of 12-deoxy-13-acylphorbol [419].

The full molecular mechanisms of the Berenblum tumor-promoting assays are still unclear. Induction of cellular proliferation was first considered as a plausible mechanistic hypothesis. However, irritant skin mitogens capable of inducing epidermal hyperproliferation as effectively as **2** but devoid of any tumor-promoting properties were discovered, as exemplified by the antibiotic calcimycin (ionophore A23187) [419]. In addition, structure-activity relationships studies in phorbol esters showed clear orthogonal relationships between irritancy and promoting activity, as exemplified by the introduction of unsaturations on the 12-acyl moiety of **2**. This showed a marked diminution of the promoting activity, but hardly any on the hyperproliferative capacity, with a complete analysis being reported for the 12-retinoyl analogue of **2**, a compound devoid of promoting activity but still as mitogenic as **2** [419]. In the Berenblum assay, a single application of **2** is unable to promote the carcinogenic activity of **672**, but when the study was continued with its retinyl analogue, sarcomas developed on the skin of mice, even though the retinyl analogue was devoid of promoting activity [419]. Tumor promotion became an entangled area of research, and the link between phorbol esters, tumor promotion and cancer started to unravel. To rationalize these observations, a two-stage theory was developed also for tumor promotion, which was considered the result of a mechanistically unknown priming effect of **2**, next magnified by the mitogenic activity of a retinyl analogue. Promoters were then sorted out into first-stage promoters, positive in the Berenblum assay, and second-stage promoters, negative in the Berenblum assay but capable of sustaining the activity of the first-stage promoters [419].

In addition to the uncertainties on its mechanism, the translational relevance of the Berenblum tumor-promotion assay has also been questioned [420], since it is species-specific to mice, and especially to the breed SENCAR, with rats and hamsters being far less sensitive, and guinea-pigs refractory just like mini-pigs and the mouse strain C57BL/6 [416]. In addition, differences became apparent between the type of tumors developed in the complete carcinogenesis assay and those associated with the tumor promotion assay, while no evidence for the tumor-initiating properties of **2** was observed [416].

Since the presence of trace amounts of carcinogenic compounds is unavoidable in the environment, the implications of the Berenblum tumor-promoting assay raised considerable interest in the cancer community. The human exposure to promoters could, in principle, play a major role in the etiology of human cancer, and since the activity of promoters, unlike that of inducers, can be reversed, the study of tumor promotion was perceived as a major opportunity for the prevention of cancer [418]. Due to the long time required (>20 weeks), faster alternatives to the Berenblum tumor-promotion assay were developed, with, however, variable degrees of positive results being obtained from these different assays used [416]. The ear edema test in rodents (see Sect. 6.3), the induction of ornithine decarboxylase in mouse skin, and the expression of Epstein-Barr virus in lymphoblastoid cells carrying the relevant genome emerged as the most popular proxies for the Berenblum croton oil assay [416]. The ear erythema assay is essentially focused on skin irritancy, and, while all tumor promoters are positive in this assay, the reverse is not true (see Sect. 6.3). These proxy-assays for tumor promotion were also utilized successfully for the detection of phorboid esters in plant materials and in the environment [421].

Despite a very large number of studies spanning five decades, no clear-cut correlation between human cancers and the exposure to promoters has been documented. Eventually, interest in this area gradually declined, and there is now a general consensus that non-genotoxic carcinogenesis in the mouse skin is not relevant to humans [420], and that the Berenblum tumor promotion assay could simply be the result of a special skin sensitivity of certain mice strains to the repeated application of irritant compounds, or even to the down-regulation of protein kinase C expression induced by phorbol esters. Remarkably, rather than on tumor promotion, the focus of research on croton oil is now in the realm of facial rejuvention: over the past two decades, all review articles on this oil were, paradoxically, in the realm of chemical peels rather than on cancer research (see Sect. 6.4).

The Croton Oil-Induced Rodent Ear Edema Test

This test is a popular in vivo model to investigate topical acute anti-inflammatory activity. It is simple, rapid and, at least in its current versions, only requires a small amount of test compound. Under ether anesthesia, a 2% solution of croton oil in a mixed solvent (e.g. ether-ethanol-pyridine) [422] is applied on one ear of a rodent (mice or rats). On the contralateral ear, the same croton oil solution containing the compound to be assayed is applied. Croton oil causes ear inflammation and a measurable increase of thickness, clearly detectable already one hour after its application, and anti-inflammatory activity is evidence by a decrease of ear thickness, within a peak of activity within four hours from the application. The effect is quantified after sacrifying the animal and comparatively weighing the control and the treated ears [422] or a plug taken from them [423].

Alternatively, ear thickness can be measured by caliper, but great care has to be taken not to squeeze edema fluid out of the tissue [424]. Activity is expressed in terms of increased weight of the treated ear as percentage of the weight of the contralateral control ear treated with croton oil alone. Being based on a relative measurement, the ear edema assay is not affected by the variability of composition of croton oil and its content of **2**.

The test is based on the PKC-dependent activation by tigliane esters of an inflammatory cascade, which eventually results in the stimulation of vascular permeability, vasodilation, and activation of phospholipase-A2 (PLA2) [422]. An inflammatory "soup" is generated, which includes histamine, serotonin, and the inflammatory eicosanoids produced by cyclooxygenase (COX) and 5-lipoxygenase (5-LOX) enzymes. Inhibitors of these enzymes as well as corticosteroids are active as anti-inflammatory agents in the assay [422–424]. While all phorbol esters active as tumor promoters are also active in the rodent ear edema assay, the reverse is not true, and no convincing rationale has so far been provided for this observation.

The Brine Shrimp Toxicity Assay

Phorbol esters show very high toxicity toward the larvae of brine shrimp (*Artemia salina*) [425], an aquatic crustacean native to salt lakes in the Mediterranean region and are popular organisms for testing the toxicity of chemical compounds [98]. Brine shrimp cysts are available year-round, and the assay can be carried out as a rapid in vivo model with mg amounts of a compound or an extract under conditions typical of a chemical laboratory [426]. A strict correlation was found between activity in assays of tumor promotion assays and of brine shrimp lethality, suggesting that this simple in vivo assay deserves attention as a general biological detection means for cocarcinogenic (a term no longer used)) phorbol esters [425] in euphorbiaceous plants. However, with the benefit of hindsight, given the sensitivity of the assay also to many antibacterial-, antifungal- and cytotoxic compounds, the value of the screening is limited to plants suspected to contain phorbol esters.

Animal Pharmacokinetic Studies

Studies on the metabolism of tigliane esters are only available for phorbol 12-myristate 13-acetate (**2**), and were not carried out after oral administration, but after topical treatment [427], skin tissues grown under laboratory conditions [428], or microsomal preparations [429]. In all these systems, the major metabolic pathway of **2** is hydrolytic, with stepwise loss of the ester groups bound at the C-12 and C-13 hydroxy groups and biologically inactivation of PKC-related activities [427]. Reduction of the C-3 carbonyl was also identified as a minor pathway in murine skin [427], while no significant involvement of P450 enzymes was observed. Similar results were also observed in cellular studies on liver microsomes [429], which also produced a significant relationship between the rate of hydrolysis and the nature of the acyl decoration, with the ester group at the 12-hydroxy being more reactive than that at the 13-hydroxy, and certain acyl groups being resistant to hydrolysis (decanoate, retinoate) [429]. Marked differences between fibroblasts of different animal origin were also observed [429]. In in vivo studies based on the application of **2** on the back skin of mice, an additional metabolic pathway was identified, namely, the esterification of the 20-hydroxy group with long-chain fatty acids [427]. There is preliminary evidence of a strong variation between species and tissues in their capacity to metabolize and deactivate **2**, and these differences could, in principle, explain the range of sensitivity in tumor-promoting assays observed in various animal species. Thus, a **2**-hydrolyzing esterase identified in mouse liver cytosol was found to be lacking in the mouse skin, but was present in the skin of hamsters, a species almost insensitive to the tumor-promoting activity of **2** [428]. On the other hand, no hydrolysis of **2** could be observed in hamster skin in vivo, questioning the pharmacokinetic explanation for the differential sensitivity to tumor-promotion within laboratory animals. A **2**-hydrolyzing esterase was show to be present in serum of mice, rats, guinea pigs, rabbits and goats, and **2**-type activity could also be detected in the plasma of leukemia patients involved in a phase 1 study on this compound as anticancer drug [430] (see 6.4).

Overall, the absorption, distribution, metabolism or excretion of **2** remain poorly known, and significant differences could exist in analogues having unsaturated or poly-unsaturated acyl moieties, since hydroxylation of the allylic positions or epoxidation of the olefin bonds by P450 enzymes could take place. In any case, no significant evidence seems to exist to show that insensitivity to tumor-promoting assays is related to the ability to detoxify tigliane esters by refractory animal species. Since no evidence was found that rumen microbes can degrade tigliane esters [431], no difference in sensitivity to the toxicity of these compounds probably exists between ruminants and monogastric animals.

Clinical Veterinary Studies

The type-C epoxytigliane ester tigilanol tiglate (**3**) has been developed as a topical veterinary anticancer agent, and is the first and so far the only tigliane ester used in medicine as a drug. The mechanism of action of **3** is based on the induction of a preliminary acute inflammatory response at the treatment site, followed by therecruitment of immune cells and the collapse of the tumor vascular system [432]. Being basically tumor agnostic, **3** has the potential to find use in different tumor types not only in companion animals, but also in humans [432], making its veterinary development worth describing.

A case study on the topical anticancer activity of **3** was first reported on the lesions associated with the canine papillomavirus type 4 (CPV-4) in three dogs of the Hungarian Vizsla breed. No standard of care is available for managing this condition, and topical imiquimod, an immunomodualtory antiviral drug, was ineffective. The size and the number of the lesions were greatly reduced by two topical applications of a gel containing compound **3** [433]. Spurred by these results, a dose de-escalation activity study was then carried out on canine mast cell tumors, the most common cutaneous neoplasm in dogs. The current management of this cancer is surgical resection, a treatment associated with a considerable degree of recurrence. A high efficacy was observed at a 1.0 mg/cm^3 concentration of **3**, while only low grade, mild and transient side effects were observed [434]. These encouraging results provided a basis for randomized controlled studies to evaluate the efficacy, safety, and recurrency rate of tigilanol tiglate (**3**) as a local intratumoral agent for the management of mast cell tumors in dogs [435, 436].

A first study was carried out with a formulation containing 1 mg/cm^3 of tigilanol tiglate (**3**) injected intratumorally at a dosage based on tumor volume [435]. The first phase of the study involved 81 dogs in the treatment cohort, and 42 dogs in the control cohort. After a single treatment, 75% complete responses were observed after 28 days, with no recurrence in 93% of dogs after 84 days. In the second phase of the study, dogs that had shown an unsatisfactory response were treated with a second dosage of **3**, and eight additional complete responses were observed, with the overall positivity reaching 88%. Remarkably, the adverse events were mild, transient and associated with the mode of action of tigilanol tiglate [435]. A second study evaluated the durability of the response at six and twelve months after the treatment on sixty-four dogs from the previous study [436]. Recurrences were detected in 11% of these, with the rest remaining still recurrence-free one year after the treatment, showing that the response to tigilanol tiglate was durable [437]. Overall comparison of tigilanol tiglate (**3**) with other therapeutic options, mainly surgery, showed a remarkable improvement in the response rate and its durability [437, 438].

Additionally, tumor margins remaining after the treatment were smaller than those associated with surgical excision, making treatment with **2** especially valuable for tumors localized near the eyes, nose and mouth [438]. Promising activity was also observed in canine high-grade mast cell tumors, a type of malignancies associated with a poor prognosis, which are more invasive, and with higher rates of local recurrence, when compared to lower-grade mastocytomas [439].

Based on these studies, **3** was approved in 2020 by the U.S. FDA and by the EMA in Europe as a topical anticancer agent for non-metastatic dog mastocytomas. The efficacy of **3** was also observed in two horses suffering from, respectively, a fibroblastic sarcoid and a fast-growing peri-ocular squamous cell carcinoma. In both cases, hemorrhagic necrosis of the tumor was observed within 24 h from the application, followed by complete re-epithelialization within 6–16 days [433]. No recurrence was observed at 93 and 189 days for, respectively, the sarcoid and the squamous carcinoma [432]. Taken togehter, these studies have established **3** as a cinically efficacious veterinary anticancer agent, boding well for its future use in a range of animal species and in a variety of cancers.

In a further distinct series of studies, a semi-synthetic analogue of **3**, 12,13-deacyl-12,13-dihexanoyl epoxytigliane (**674**) (EBC1013), was shown to accelerate and improve wound healing in several relevant animal models (acute thermal injuries in calves and chronically infected diabetic mouse wounds), an activity related to a local stimulation of PKC, leading, possibly via the intermediacy of TRPA1 [440], to an increased expression of wound healing-associated cytokines, chemokines, and antibacterial peptides. Notably, EBC1013 also showed considerable antimicrobial activity, potently inducing biofilm disruption [441]. These results have provided a rationale to start its clinical development for acute wounds in horses [442].

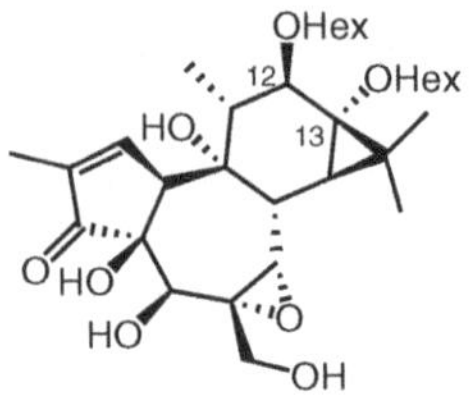

674

Animal Toxicological Studies

Due to the irritancy of tigliane esters, vertebrates do not generally feed on plants containing these compounds, with the notable exception of rhinoceroses and a few other species of African animals living in dry areas [443, 444]. The molecular basis for insensitivity to the offensive properties of these compounds is unknown, and has been the source of much speculation, and the current view associates it with the presence of specific detoxification pathways rather than mutations in protein kinase C, although this hypothesis should be worth investigation within insects. The sensitivity of insects to tigliane esters is unclear.

Locusts, cockroaches, and possibly other insects as well, express a PMA (**2**)-sensitive PKC in, respectively, their nervous system and their corpora allata. Treatment with PMA (**2**) potentiates the release of acetylcholine at nervous junctions [445] and inhibits the production of the juvenile hormone III in the corpora allata [446]. Therefore, it is generally assumed that tigliane esters are toxic to insects, although bees can feed with impunity on several euphorbiaceous plants and transfer tigliane diterpenoids to the honey [444]. Some insects feed on tigliane-containing plants, even developing their larvae into euphorbiaceous latices and toxic phorbol esters have been reported to be sequestered by some aposemantic insects [447], with a pre-eminent example being the larvae of various *Hyles* feeding on plants from the genus *Euphorbia* [447]. Tigliane esters were identified in *H. euphorbiae* [447], and a draft transcriptome of *H. euphorbiae* showed the upregulation of all four major classes of drug-metabolizing enzymes (P450 cytochromes, carboxylesterases, glutathione *S*-transferases, and lipases), providing a plausible explanation for the immunity to tigliane esters of these insects [448, 449]. On the other hand, the toxicity of tigliane esters for many crop pests is well established [450], and has provided a rationale for the use of seed pannel extracts from oleaginous euphorbiaceous plants for crop protection [451]. The idea gained momentum from the development of extensive cultivations of tigliane-containing oleaginous euphorbiaceous plants (*Jatropha curcas*, *Croton tiglium*) in the tropics, but several issues remain unresolved. Due to their toxicity, the extracts can not be applied to the edible parts of food plants, and little information is available on the kinetics of degradation of tigliane esters in soil. The limited atmospheric stability of phorbol should secure a fast degradation, but it has been reported that ether extracts from the soil around tigliane-producing plants test positive in the Epstein-Barr activation assay, a proxy for tumor-promoting activity [452]. Due to the presence of tigliane esters not only in *Jatropha* oil but also in *Jatropha* pannel [453], the possibility of contamination from bio-refinery side-products seems therefore to exist [453], and similar limitations occur for the use of pannel extracts to kill slugs and snails.

The forced consumption of oleaginous euphorbiaceous kernel meals has been associated with weight loss, damage to the gastrointestinal mucosa, and liver and kidney lesions [339]. Based on the decrease of feed intake and body gain, a no-observed-adverse-effect-level of 0.4 mg tigliane esters (TPA equivalent)/kg body weight (bw) per day was calculated in pigs [339]. Due to the paucity of data, a no-observed-adverse-effect-level could not be established for companion animals, ruminants, horses, poultry, and aquatic species. No evidence has so far been reported that rumen microorganisms can detoxify tigliane esters, and no difference in sensitivity to the exposure to these componds seems therefore to exist between ruminant monogastric animals [339]. The most well-investigated livestock intoxication by phorboids is St. George disease, a major cause of economic loss in Australia and New Zealand. This poisoning is caused by the ingestion of toxic *Pimelea* species [454]. These daphnaceous plants are rarely eaten by livestock, but they can grow mixed with palatable plants and be consumed inadvertently or eaten when pasture availability is minimal. Almost 150 different *Pimelea* species have been identified in Australia and New Zealand, and some of these have ornamental value, and have become popular also in Europe. The presence of tigliane and daphnane esters is the phytochemical hallmark of these plants. Since most studies on St. George Disease have focused on daphnane esters [454], this poisoning will be discussed in a forthcoming daphnane review in this book series, as will manchineel poisoning.

Fish are very sensitive to the toxicity of tigliane esters, and euphorbiaceous plants have been used wherever they grow for fishing purposes. Carp show a high sensitivity to phorbol esters, and a threshold level for toxicity was estimated at 15 ppm phorbol esters in the diet. Very strict limits of concentration of these compounds exist therefore in protein feeds obtained from euphorbiaceous pannels and used for fish and shrimp aquaculture [455]. Fish remain edible after poisoning from tigliane esters, and fishing with tigliane-containing plants is a veritable equivalent of hunting with curare.

6.4 Clinical Studies

The Croton Oil Peel

Most human data on exposure to tigliane esters are related to the croton oil peel, a dermatological practice that seems to swim against the current of common sense, given the ravages that croton oil produces on the skin. This protocol of long-lasting skin rejuvenation originated from a formulation of croton oil and phenol developed in the 1920s by Antoinette La Gasse, an African American, for lightening her brown skin [456]. La Gasse noticed the remarkable anti-wrinkle effects of her peel, and started a successful cosmetic practice in California. Croton oil had already been used by Ferdinand Ritter von Hebra (1816–1880), the Austrian founder of modern dermatology, to treat freckles, severe acne, and skin pathologies that are not esthetic, and military surgeons in the First World War had noticed its beneficial effects on wound scars. Nevertheless, its use for non-medical cosmetic purposes apparently was not yet popular or documented [457].

La Gasse was eventually arrested in 1939 for the unlicensed practice of medicine, and when she died of cancer in 1952, the formula was inherited by her collaborator, Cora Galenti (1898–1993), a Sicilian clothing designer working in Hollywood for the Fox and Paramount studies. Galenti improved the formula and made it popular with movie celebrities, with Marlene Dietrich being one of the few admitting having her wrinkles ironed out by chemical peel [458]. Galenti built a veritable empire based on her secret deep-peel skin rejuvenating formula, also treating for free scarred war veterans and teenagers with severe acne. However, when Galenti's formula fell under the scrutiny of mainstream medicine, the presence of croton oil and phenol made it immediately shift from the realm of cosmetics to that of medicine, causing Galenti to be fined for practicing medicine without a license. In 1959, Galenti moved her business to Las Vegas, but eventually she ended up into jail for mailing flyers with false testimonials that her treatment was "painless" [456]. The composition of Galenti's peel raised eyebrows among clinicians, but its remarkable activity attracted a considerable and curious interest within the plastic surgeon community. The peel was incorporated into mainstream practice by the surgeon Thomas Baker, who in 1961 published a first detailed account of its composition, standardizing the formula as 47.5% phenol, 1.2% croton oil, and 2% surfactant [459].

At that time, the active ingredient was considered phenol, and the peel was accordingly named "phenol peel", also because of the sinister reputation of croton oil as a tumor promoter. Later studies showed that the active ingredient was indeed croton oil and not phenol, which behaves as a carrier of croton oil phorbol esters and also hinders them from penetrating too deep in the dermis, and so the name was changed accordingly [459]. The treatment dramatically improves sun-damaged skin and removes age-related depth wrinkles. During exposure to the peel, the surface layers of the skin are gradually damaged as the oil moves through the epidermis to the papillary- and reticular dermis. As a result of this diffuse damage, a copious production of collagen, both of type-1 and type-3, and elastic fibers is stimulated, and the damaged skin is replaced by a healthy one [460]. At least with the original Baker formulation and its more concentrated Baker-Gordon modification, loss of the capacity to tan or a significant hypopygmentation were not rare, making only people with fair skin eligible for the use of the peel [456]. In addition, the destructive phase of the peeling process lasted from a minimum of two weeks to several months, during which time special skin bandages and ad hoc skin care were necessary [456].

The popularity of the croton oil peel was boosted in 2000 by the development of a novel formulation by the plastic surgeon Gregory Hetter, who, in four notes, critically analyzed the formula of the peel and its practical use, eventually decreasing the concentration of phenol, modulating that of croton oil according to skin type, age, and clinical need, and optimizing the surfactant necessary to produce a stable dispersion of croton oil in water [456, 459, 461, 462]. By modulating the peel strength, the application became possible to all ages and to many skin types, heralding a new era in chemical peeling [463]. Considerable ability is required, in any case, from the surgeon, who must judge the amount of oil used [461], while skin devastation, temporal facial disfiguration, burning, swelling and blistering have not apparently proved too intimidating, since the peeling has been used on thousands (if not millions!) of people.

Nowadays, croton oil peel remains the standard against which all the other skin rejuvenating processes are judged, superior in terms of results and durability to laser treatment, and with outstanding cost-effectiveness [463].

Surprisingly, the rich literature on the side-effects of croton oil peel, which includes phenol-related cardiac arrhythmias, is focused on the acute phase of the protocol, and follow-up studies to identify delayed toxicity outside waxy-skin appearance, hypopigmentation, and scarring, apparently have never been carried out or published [464]. Little information is also available on the absorption of phorbol esters from the topical application of croton oil, but the possibility of inducing herpetic outbreak and the requirement of antiviral prophylaxis suggest a certain degree of penetration [463]. The croton oil peel has been used extensively especially in the U.S. and Brazil, and scrutiny of the long-term toxicity associated with the treatment could afford interesting information on the safety of tigliane-based drugs, especially when administered by intratumoral injection. The claim that the croton oil peel can eliminate certain pre-cancerous lesions [463] needs careful scrutiny and clinical validation.

The mechanism of the skin-rejuvenating activity of croton oil is poorly known. It is the result of a disruption of tissue architecture followed by its restoration, a process basically similar to wound healing but which, however, occurs without forming fibrotic scars. The ion channel TRPA1, strongly expressed in cutaneous peripheral sensory neurons, is critically involved in scar-free skin regeneration, and it is stimulated greatly by activated oxygen species [464]. Phorbol esters, by potently inducing the generation of activated oxygen species could therefore induce a TRPA1-mediated neuroimmune skin-regenerating cascade, stimulating the local production of interleukin-23 in dermal dendritic cells. In a murine model of wound-healing, this in turn activates $\gamma\delta$ T cells, which are known to promote tissue regeneration, and a similar T cell population exists in human skin [464]. Skin wounds are unable to activate this neuroimmune pathway, and this could be the reason why they heal with scarring [464].

Anticancer Clinical Studies

Phorbol myristate acetate (**2**) was investigated as a chemotherapeutic agent both as a stand-alone drug and as part of anticancer drug cocktails [430, 465–467]. These studies aimed at translating into a validated clinical activity the extraordinarily (sub-nanomolar to one-digit nanomolar) potency of this phorbol ester in the differentiation-induced inhibition of proliferation of cultured human promyelocytic leukemia cells as well as peripheral leukemia cells from patients with acute myelocytic leukemia [468]. Given the ultrapotency of **2**, it was hoped that the intravenous administration of the compound could achieve bioactive concentrations without eliciting severe general toxicity. A first study was carried out in twelve patients with refractory myelocytic leukemia, using **2** alone or in combination with vitamin D_3 and low dosages of cytosine arabinoside (ara-C). A decreased number of myeloblasts in blood and the general tolerability were the end-points of this study [430]. The regimen of intravenous administration was varied from 0.25 to 2 mg **2** per week, and the treatment lasted from two to eleven weeks [430].

Although all patients were bedridden and severely ill, five of them could leave the hospital and return to a normal lifestyle, experiencing a temporary complete remission of the disease. For an ara-C resistant patient administered 1 mg **2** and 40 mg ara-C once a week for three weeks, and vitamin D_3 twice a week for two weeks, the remission lasted for eight months. In other cases, the remission was shorter, in the range of one to five months. In two cases, the hematological end-point was reached, but the overall recovery was only partial. In the two cases where **2** was administered at a lower dosage (0.25–0.5 mg), but with only 2–3 days interval, the side-effects were marked, and the treatment had to be discontinued [465]. Dyspnea was observed at the higher dosages as the major side-effect, possibly due to pulmonary irritation [465]. Overall, the responses were of relatively short duration, but highly significant in respect to the refractoriness of the disease states. In addition, the lack of bone marrow hypoplasia associated with the treatment strongly suggested that lymphocyte differentiation was at the basis of the observed responses. Taken together, the results of this first clinical study showed that **2**, either alone or in combination, can elicit clinically beneficial effects on cancer patients who are refractory to conventional chemotherapy [465].

A follow-up dose-escalation study was carried out in thirty-five leukemia patients to establish an optimum dosing regimen for **2**, starting with 0.063 mg/m^2 (ca. 0.10 mg for a patient of ca. 70 kg) intravenous dose administration on days 1–5 and 8–12, followed by a two-week rest [466]. The major side-effects identified were hematuria, seizures, syncope, and hypotension. Coagulation issues were not reported, despite the powerful thrombotic activity of **2** (78). Transient fatigue, mild dyspnea, fever, rigor, and muscular aches were generally experienced after the infusion. However, a complete response at well-tolerated dosages was only observed in one patient.

A third clinical study [467] combined **2** with the kinase inhibitor imatinib mesylate in twelve chronic myeloid leukemia patients resistant to a standard-dosage (400 mg/day) imatinib. The phorbol diester **2** was administered intravenously at a dosage identified as well-tolerated in the previous study (0.063 mg/m^2) on day 1 and day 7 of the cycle, and imatinib daily up to hematological recovery, for a maximum of 70 days. The cycle was repeated after a one-week rest, and the end-point of the study was the partial or complete disappearance of all signs and symptoms related to disease (normalization of peripheral counts and a normal bone marrow morphology with $\leq 5\%$ marrow blasts). Overall, four patients achieved a complete hematological response and four a partial one, with survival ranging from 5 to 27 months [467].

The three clinical studies carried out in 1998–2007 with **2** have shown that the compound is relatively well-tolerated at active dosages, and can induce clinically significant responses in chemotherapy-resistant leukemia patients, both alone and in association with other chemotherapeutic drugs, as shown for imatinib. No clinical follow-up work, however, was done after the completion of these studies. An additional phase II clinical study of **2** in combination with dexamethasone and choline magnesium trisalicylate for the treatment of patients with relapsed or refractory acute myelogenous leukemia was terminated early due to a shortage of drug-grade **2** [469].

More recently, a non-randomized, single arm, phase-I dose-escalation study was carried out on the epoxytigliane tigilanol tiglate (**3**) to evaluate safety, tolerability, preliminary efficacy, and pharmacokinetics of its intratumoral injection in twenty-two patients with nine distinct solid tumors [470]. Dose escalation started with 0.06 mg/m^2 injected in tumors for which the volume was at least twice that of the injection, and the amount injected could be increased six-fold without reaching limiting toxicity. In addition to this dose-escalation cohort, a second cohort for efficacy recruited patients with smaller tumors, who received a dose that was dependent on the size of their tumor. Upper airway obstruction and septicemia were the most serious adverse events observed, and, overall, 160 adverse events were recorded, all manageable without any fatalities. In all tumor types, a reaction occurred at the lowest dose, and six treatment responses were documented, of which four were completed [139]. This study has provided a solid rationale for the further development of tigilanol tiglate (**3**) as a topical anticancer agent.

Wound-Healing Studies

The croton oil peel vividly demonstrates the ability of tigliane esters to induce skin regeneration and, potentially, to induce a neuroimmune-mediated, scar-free wound healing [463] In addition, keratinocytes express thermo-TRPs like TRPV1, TRPV2, and TRPV4, which can be activated by phorbol esters in an indirect way, via the PKC-dependent generation of activated oxygen species [389]. In addition to skin-regenerating properties, a potent antibacterial and anti-biofilm activity was discovered for the tigilanol tiglate analogue **674**, which, after animal and veterinary studies, is now entering phase I clinical studies for the management of diabetic leg ulcers [442].

6.5 Dietary Exposure to Tigliane Esters

Given the skin-irritant properties of phorbol esters, their presence in food makes it unpalatable for humans, although camels and rhinoceroses can feed with impunity on plants containing this type of compound, and some insect larvae even develop in latices laden with irritant phorboids [471]. Nevertheless, the risk of chronic human exposure to traces of phorbol esters through the food chain exists in principle, due to fodder contamination from phorbol-ester containing plants [472] or from animal feed obtained from byproducts of biodiesel production from oleaginous euphorbiaceous seeds [472]. Most studies in this area focus on the oil pannel from *Jatropha*, where the concentration of phorbol esters can be reduced to levels that do not compromise animal palatability, from 0.6–3.7 g/kg, to less than 3 mg/kg, an amount considered safe for animal consumption [339]. Nevertheless, given the lipophilic nature of phorbol esters and the very limited information existing on their metabolism, the possibility exists of accumulation and transfer through the food chain, a process for which few details are available. A human exposure scenario was calculated using a 50% transfer rate from feed to milk and a daily intake of 1 μg *Jatropha* phorbol (actually 16-hydroxyphorbol) esters per kg per day, concluding that no firm conclusions on human risk could be drawn [339].

In the context of chronic exposure to tigliane esters, considerable attention has been given to a possible correlation between the high incidence of esophageal cancer in the island of Curaçao and the local popular use of a tea from the leaves of *Croton flavens* L. The plant, locally known as "welensali", contains 16-hydroxyphorbol esters active in assays of tumor promotion [124, 473]. These compounds have been reported to be detectable in welensali teas in concentrations active in the mouse-ear irritancy assay, and possibly sufficient to maintain a state of chronic irritation that can favor the occurrence of cancer [473]. On the other hand, welensali has a very rich phytochemistry [474], and its leaves showed no cancer-inducing activity in animal experiments [475]. None of the more recent investigations on the leaves of *C. flavens* have reported the occurrence of phorbol esters, and it is possible that the biomass analyzed by Hecker and found to contain 16-hydroxyphorbol esters [124] was not actually *C. flavens*, or, alternatively, was from a plant part different from the leaves. Remarkably, based on the analysis of twenty-seven retrieved articles, the draft of the EFSA Compendium of Botanicals, classified *C. flavens* as "approved" [476], either overlooking the Hecker studies or not considering them reliable.

Many tigliane-containing species are excellent melliferous plants. Since the intracellular latex-secretion and the extracellular nectar-secretion are anatomically connected [477], the possibility exists of contamination of honey by tigliane esters. This, indeed, has been reported in Noors honeys from South Africa [478] and in honeys from the Caucasus [479], which show a distinctive pungent and bitter note. The toxic properties of honeys from the Caucasus has been associated traditionally with the presence of other diterpenoids (grayanotoxins from *Rhododendron* species) [480], but the potential presence of tigliane esters should also be taken into consideration. Noors honeys are hardly palatable, and are not commercialized or consumed significantly, but rather used by beekeepers to feed bees, which are not poisoned by them. The highly prized Moroccan Daghmous honey ("miel d'euphorbe") is devoid of a significant bitter or burning taste, rather being described as having a sweet and mild taste, with only a subtle note of slightly acridity [481]. This observation is surprising, since one of its major sources is *Euphorbia resinifera* Berg., a plant containing, in addition to tigliane esters, also resiniferatoxin (**10**), a compound more potent than **2** in the mouse ear erythema assay, and of capsaicin in terms of pungency [482]. This observation deserves a systematic investigation, since phorboids could be detoxified by certain pollinating bees and not by others.

Phorbol ester-containing oils inadvertitedly can enter the food chain, as reported in 1937 in Hamburg, when a barrel of tung oil (the oil from the seeds of *Aleurites fordii* Hemsl.), used to make water-resistant boat varnishes, was sold by mistake to a baker as rape oil. He then used it to make pancakes, causing diarrhea and vomiting in 190 people, without, however, any fatalities [483].

6.6 Accidental Exposure to Tigliane Esters

With the notable exception of the fruits of the Mancinella tree (*Hippomane mancinella* L.), which is initially sweet and pleasant (but then extraordinarily irritant [484]), the burning taste of tigliane esters makes it relatively uncommon for the accidental ingestion of their plant source in amounts sufficient to induce systemic toxicity. When this occurs, the painful inflammation of the throat deters further consumption, while vomiting and severe diarrhea limit systematic absorption [485]. Most cases reported refer to children who contaminated their fingers with the latex of spurges, and then sucked them. Despite the severity of pain in the oral cavity, this poisoning does not generally lead to systemic effects, but a case was reported in the U.K., where the consumption of *Euphorbia helioscopia* L. by a child was associated with pulmonary edema, convulsions, and eventually death [485]. Much more common is, conversely, skin toxicity by contact with the latex of tigliane-containing plants, especially when gardening, since some tigliane-containing plants can be found within popular ornamentals as well as in weeds [485]. In addition, unlike the immediate oral response to miglianes, the onset of the skin reaction is often delayed, sometimes by 2–8 h after the contact, with the possibility of cumulated exposure. This causes reddening and formation of edematous swelling, vescicles, and pustules. Of particular severity are the effects on the eye, which include intense pain, decrease of visual acuity, and, in severe cases, destruction of the corneal epithelium, and necrosis of the cornea and loss of vision [486]. Contrary to a widespread belief, the latex of poinsettia (*Euphorbia pulcherrima* Willd. ex Klotzsch.), a New World species, is neither toxic nor irritant, and the plant does not contain tigliane esters, for which their occurrence is the hallmark of Old World Euphorbias [487]. Irritant tigliane esters are, however, present in many popular African succulents, including *E. resinifera* Berg. and *E. officinalis* L. [482]. Manchineel (*Hippomanae mancinella* L.), the most dangerous plant in the world according to the Guiness Book of Records [488], contains tigliane esters [489]. However, the most spectacular effects of its toxicity, like the almost immediate formation of blisters and burns upon contact with its sap, do not fit well with the kinetics of skin toxicity of phorbol esters, and the co-occurring daphnane esters as well as other constituents, could also be involved.

6.7 Exposure to Tigliane Esters Related to Folk Medicine

Plants containing tigliane-rich latices have been used in folk medicine as a depilatory and to treat skin complaints, especially for the scar-free removal of warts and freckles [490]. Remarkably, this use has been documented by the local medical systems wherever tigliane-rich plants grow [490]. The main advantage of the phorboid latex compared to other folk-recipes for the removal of warts is the lack of scarring, but misuse can lead to dangerous consequences, especially when the latex is applied next to the eyes [491]. Over the centuries, a significant, but temporally limited, skin exposure to tigliane esters has taken place, but it is difficult to associate retrospectively any long-term effects due to this use. Euphorbias produce a white sap, and this might well be the reason for their utilization as a galactogogue in accordance with the "doctrine of signatures" [492], a use documented, however, mainly for species devoid of phorboids.

6.8 Summary of Biological Studies

Most living forms, from prokaryotes to primates, are sensitive to the activity of tigliane esters, but, despite the number of studies, important gaps remain in our knowledge on the biological profile of these compounds. Some aspects of the interaction of phorbol esters with protein kinase C are still unclear, and the involvement of the hydroxy group at C-4 or that at C-9 is not definitive, rather implying potential druggable differences between the solid state and the solution conformation of protein kinase C, for which its conformational mobility, and therefore plasticity in ligand binding, has emerged clearly in solution studies. There is also a surprising paucity of information on the ecological role of phorbol esters, and a better knowledge on the molecular details of the resistance to their activity by some insects and, apparently, by some vertebrates as well, could provide interesting information on the regulation of protein kinase C activity and/or the mechanisms of detoxification of these compounds. Most importantlty, the clinical development of tigilanol tiglate (**3**) as an anticancer drug, both as a standalone agent and in combination with immunological agents such as pembrolizumab (Keytruda), could open new avenues for cancer treatment at the interface between classic cytotoxicity and immunological chemotherapy [493].

7 Conclusions

For long time, in the discipline of phytochemistry it was only possible to describe what was there, treating natural products in a descriptive kind of way, like plants, insects and other products of Nature's bounty. Natural products chemistry is, indeed,

Plate 9 Janus, the Roman God of duality, was represented with two faces looking in opposite directions, well exemplifying the dual biological profile of phorbol esters. Photograph of artwork phographed in Emilia Romana courtesy Kizel

imbibed with the culture of natural history, and the plant garden of chemical species was explored not unlike naturalists did with biological species, with a difference being that, unlike the descriptive work of naturalists, the contributions of chemists have required several decades, if not more than a century, of structure elucidation work. Enormous progress in both separation technology and spectroscopy have now made it possible to go beyond the sheer descriptive aspect of natural products, addressing issues like biosynthesis, molecular mechanism of action, and ecological roles. This has placed phytochemists in a unique position to widen the horizons of our molecular knowledge of Nature. We can now characterize compounds present in a plant at concentration levels close to molecular noise, and these compounds will surely not only expand the biological space of natural products, but also provide important biogenetic clues for genomic studies aimed at the identification of the molecular basis for the production of secondary metabolites.

Tiglianes are one of the classes of natural products expected to benefit considerably from these advancements, owing to the incomplete level of our understanding of their biosynthesis and raison d'être, and to the complexity of their biological profiles of activity. Tigliane esters are not good or bad, but both—reminiscent of the ancient Roman god Janus (Plate 9). For a long time, our attention was focused only on the dark side of their bioactivity, that of skin irritancy and toxicity, but the recent development of tigilanol tiglate as a veterinary anticancer agent suggests that the time has come to explore also the other side of their biological profile, taking advantage of their well-documented ecological and ethnopharmacological footprints.

Acknowledgements Schiller claimed that "Wenn die Könige bauen, haben die Kärrner zu tun" (When kings build, drudges have something to do). We hope our "tun" gives due recognition to the "Könige" on whose shoulders we have climbed to review the field of tigliane diterpenoids (Hecker, Crombie, Evans, Jakupovic, Wender, Baran ...). In addition, we are very grateful to Prof. Jeroen Dickshat (University of Bonn) for reading and commenting on several sections of this contribution, but for any errors and inaccuracies the authors take all responsibility. We are very grateful to Heinz Falk, whose help, suggestions, and support went well beyond his editorial duties. Finally, G.A. would like to dedicate this chapter to the memory of Jasmin Jakupovic, whose premature death meant the loss not only of a scientific model, but also of a sincere friend.

References

1. Woodward RB, Cava MP, Ollis WD, Hunger A, Daeniker HU, Schenker K (1963) The total synthesis of strychnine. Tetrahedron 19:247
2. Böhm R, diary, page 309, dated Dec. 8, 1923 (mentioned in the PhD dissertation of R. v. Wolffersdorff, Leipzig 1929)
3. Pettersen RC, Ferguson G, Crombie L, Games ML, Pointer DJ (1967) The structure and stereochemistry of phorbol, diterpene parent of co-carcinogens of Croton oil. Chem Commun 4:716–717
4. Hecker E, Bartsch H, Bresch H, Gschwendt M, Härle A, Kreibich G, Kubinyi H, Schairer HU, v. Szczepanski C, Thielmann HW (1967) Structure and stereochemistry of the tetracyclic diterpene phorbol from *Croton tiglium* L. Tetrahedron Lett 33:3165
5. Hecker E (1968) Cocarcinogenic principles from the seed oil of *Croton tiglium* and from other Euphorbiaceae. Cancer Res 28:2338
6. Newton AC, Brognard J (2017) Reversing the paradigm: protein kinase C as a tumor suppressor. Trends Pharmacol Sci 38:438
7. Nishizuka K (1984) The role of protein kinase C in cell surface signal transduction and tumour promotion. Nature 308:693
8. Cullen JK, Boyle GM, Yap P-Y, Elmlinger S, Simmons JL, Broit N, Johns J, Ferguson B, Maslovskaya LA, Savchenko AI, Mirzayans PM, Porzelle A, Bernhardt PV, Gordon VA, Reddell PW, Pagani A, Appendino G, Parsons PG, Williams CM (2021) Activation of PKC supports the anticancer activity of tigilanol tiglate and related epoxytiglianes. Sci Rep 11:207
9. Juan JC, Kartika DA, Wu TY, Hin T-Y (2011) Biodiesel production from *Jatropha* oil by catalytic and non-catalytic approaches: an overview. Bioresour Technol 102:452
10. Devappa RK, Makkar HPS, Becker R (2010) Optimization of conditions for the extraction of phorbol esters from *Jatropha* oil. Biomass Bioenergy 34:1125
11. Pal PK, Nandi MK, Anc SNK (2014) Detoxification of *Croton tiglium* L. seeds by Ayurvedic process of Śodhana. Anc Sci Life 33:157
12. Madaus G (1938) Lehrbuch der biologischen Heilmittel. Band II, Olms, Hildesheim, Germany, p 1129
13. Karrer P, Weber F, van Slooten J (1925) Über Toxine II. Zur Kenntnis des Crotins. Helv Chim Acta 8:384
14. Gaillard Y, Regenstreif P, Fanton L (2014) Modern toxic antipersonnel projectiles. Am J Forensic Med Pathol 35:258
15. Meyer-Bertenrath JG (1969) 150 years of croton oil research. Experientia 25:1
16. Pelletier J, Caventou JB (1829) J Pharm 15:514
17. Buckingham J (2004) Chasing the molecule. Sutton Publishing, Stroud, UK, p36
18. Böhm R, Flaschenträger B (1930) Über den Giftstoff im Krotonöl. Arch Exptl Pathol Pharmakol 157:115

19. Flaschenträger B, von Falkenhausen FF (1934) Über den Giftstoff im Krotonöl. II. Zur Konstitution von Krotophorbolon. Ann Chem 514:252
20. Berenblum I (1941) The cocarcinogenic action of croton resin. Cancer Res 1:44
21. Rasmussen C (1999) 1950s Face peel inventor hit legal wrinkles. Los Angeles Times
22. Clarke E, Hecker E (1965) On the active principles of Croton oil. Z Krebsforsch 67:192
23. Kauffmann T, Neumann H, Lenhardt K (1959) Zur Konstitution des Phorbols, I. Über die reduzierende Gruppe des Phorbols. Chem Ber 92:1715
24. Kauffmann T, Eisinger A, Jasching W, Lenhardt K (1959) Zur Konstitution des Phorbols, II. Über die α-Glykolgruppe des Phorbols. Chem Ber 92:1727
25. Arroyo ER, Holcomb J (1965) Structural studies of an active principle from *Croton tiglium* L. J Med Chem 8:672
26. Hecker E, Kubinyi H, von Szczepanski C, Hárle E, Bresch H (1965) Phorbol, ein tetracyclisches Diterpen aus Crotonöl. Tetrahedron Lett 6(23), 1837
27. Hecker E, Kubinyi H, Bresch H, v. Szczepanski C (1966) The chemical structure of a cocarcinogen and of phorbol isolated from croton oil. J Med Chem 9:246
28. Pattenden G (2001) Leslie Crombie. Biog Mems Fell R Soc Lond 47:125
29. Hoppe W, Brandl F, Strell I, Röhrl M, Gassmann I, Hecker E, Bartsch H, Kreibich G, v. Szczepanski C (1967) Röntgenstrukturanalyse des Neophorbols. Angew Chem 79:824
30. Crombie L, Games ML, Pointer DJ (1968) Chemistry and structure of phorbol, the diterpene parent of the co-carcinogens of croton oil. J Chem Soc C Org:1347
31. Hecker E (1967) Phorbol esters from croton oil chemical nature and biological activities. Science Nat 54:282
32. Hecker E, Schmidt R (1974) Phorbolesters—the irritant and cocarcinogens of *Croton tiglium* L. Progr Chem Org Nat Prod 31:283
33. Evans FJ, Taylor SE (1983) Pro-inflammatory, tumor promoting and anti-tumour diterpenes of the plant families Euphorbiaceae and Thymelaeaceace. Progr Chem Org Nat Prod 44:17
34. Ogbourne SM, Parsons PG (2014) The value of Nature's natural products library for the discovery of new chemical entities: the discovery of ingenol mebutate. Fitoterapia 98:36
35. Hergenhahn M, Adolf W, Hecker E (1975) Resiniferatoxin and other esters of novel polyfunctional diterpenes from *Euphorbia resinifera* and *unispina*. Tetrahedron Lett 16(20):1595
36. Chianese G, Amin HIM, Maioli C, Reddell P, Parsons P, Cullen J, Johns J, Handoko H, Boyle G, Appendino G, Taglialatela-Scafati O, Gaeta S (2022) Cryptic epoxytiglianes from the kernels of the blushwood tree (*Fontainea picrosperma*). J Nat Prod 85:1959
37. Pieters LA, Vlietinck J (1986) Rotation locular countercurrent chromatography and quantitative ^{1}H-nuclear magnetic resonance spectroscopy of the phorbol ester constituents of croton oil. Planta Med 52(6):466
38. Hirota M, Ohigashi H, Koshimizu K (1979) Piscicidal constituents and related diterpene esters from *Aleurites fordii.* Agric Biol Chem 43:2523
39. Saputera M, Atikaha TA (2014) Extraction and transesterification of *Croton tiglium* oil seeds from central Kalimantan, Indonesia, as an alternative biodiesel raw materials. Asian J Appl Sci 7:140
40. https://www.sigmaaldrich.com/IT/it/product/sigma/c6719. Accessed on 26 Oct 2023
41. Flaschenträger B, Wigner G (1942) The poison in croton oil. The isolation of croton resin, the oil and phorbol from croton oil by alcoholysis. Helv Chim Acta 25:569
42. Pagani A, Gaeta S, Savchenko AI, Williams CM, Appendino G (2017) An improved preparation of phorbol from croton oil. Beilstein J Org Chem 13:1361
43. Fan JP, He BT, Gao Y, Xie CF, Chen HP, Peng HL (2023) Biphasic alcoholysis coupled with high-speed countercurrent chromatography for high performance on separating phorbol from *Croton tiglium* L. extracts. J Sep Sci 46:2200984
44. Zimmermann T, Franzyk H, Christensen SB (2018) Phorbol rearrangements. J Nat Prod 81:2134
45. Wender PA, Gentry ZO, Fanelli DJ, Luu-Nguyen QH, McAteer OD, Njoo E (2022) Practical synthesis of the therapeutic leads tigilanol tiglate and its analogues. Nature Chem 14:421

46. Sperti S, Montanaro L, Mattioli A, Testoni G, Stirpe F (1976) Inhibition of protein synthesis in vitro by crotins and ricin. Effect on the steps of peptide chain elongation. Biochem J 156:7
47. Jaspersen-Schib R (1976) Exotische Halsketten aus toxischen Samen und Früchten. Schweiz Apoth Ztg 114:391
48. Bertolini TA, Giorgione J, Harvey DF, Newton AC (2003) Protein kinase C translocation by modified phorbol esters with functionalized lipophilic regions. J Org Chem 68:5028
49. Appendino G, Carello GP, Enriù R, Jakupovic J (1995) A new oxidative rearrangement of a phorbol derivative. J Nat Prod 58:284
50. Bucheim R (1857) Virchow's Arch Pathol Anat Physiol Klin Med 17:1
51. Pérez-Jiménez J, Torres JL (2011) Analysis of nonextractable phenolic compounds in foods: the current state of the art. J Agric Food Chem 59:12713
52. Hecker E, v. Szczepanski C, Kubinyi H, Bresch H, Härle E, Schairer U, Bartsch H (1966) Über die Wirkstoffe des Crotonöls VII. Phorbol. Z Naturforsch 21b:1204
53. Fürstenberger G, Hecker E (1977) The new diterpene 4-deoxyphorbol and its highly unsaturated irritant diesters. Tetrahedron Lett 18(11):925
54. Moro Loureiro FA, Cavalcante de Oliveira AL (2021) Evaluation of the methodology for the chemical extraction of phorbol esters contained in *Jatropha curcas* oil. J Environ Anal Toxicol 11:S7
55. Buděšínský M, Šaman D (1987) Identification of acyl groups occurring in sesquiterpene lactones: proton and carbon-13 NMR study. Collect Czech Chem Commun 52:450
56. Appendino G, Belloro E, Barboni L (1997) Unusual reactivity of taxine A in an oxidation-reduction protocol. Gazz Chim Ital 127:373
57. Bresch H, Kreibich G, Kubinyi H, Schairer HU, Thielmann HW, Hecker E (1968) Über die Wirkstoffe des Crotonöls, IX. Partialsynthese von Wirkstoffen des Crotonöls. Z Naturforsch 23b:538
58. Menzel JP, Young RSE, Benfield AH, Scott JS, Wongsomboon P, Cudlman L, Cvacka J, Butler LM, Henriques ST, Poad BLJ, Blanksby SJ (2023) Ozone-enabled fatty acid discovery reveals unexpected diversity in the human lipidome. Nat Commun 14:3940
59. Wang Z, Garza S, Li X, Rahman MS, Brenna JT, Wang DH (2024) Paternò-Büchi reaction mass spectrometry enables positional assignment of polymethylene-interrupted double bonds in food-derived lipids. J Agric Food Chem 72:3180
60. Nothias LF, Nothias-Esposito M, da Silva R, Wang M, Protsyuk I, Zhang Z, Sarvepalli A, Leyssen P, Touboul D, Costa J, Paolini J, Alexandrov T, Litaudon M, Dorrestein PC (2018) Bioactivity-based molecular networking for the discovery of drug leads in natural product bioassay-guided fractionation. J Nat Prod 81:758
61. Hoppe W, Zechmeister K, Rohrl F, Brandl E, Hecker E, Kreibich H (1969) The structure determination of a solvate of phorbol, the diterpene parent of the tumour promoters from Croton oil. Tetrahedron Lett 10(8):667
62. Brandl F, Röhrl M, Zechmeister K, Hoppe W (1971) Röntgenstrukturnalysen von Neophorbol $C_{31}H_{35}O_9Br$ und Phorbol $C_{20}H_{28}O_6$. Acta Crystallogr Sect B 27:1718
63. Petersen RC, Birnbaum GI, Ferguson G, Islam KM, Sim JG (1968) X-ray investigation of several phorbol derivatives. The crystal and molecular structure of phorbol bromofuroate-chloroform solvate at −160°. J Chem Soc B 1:2
64. Nothias L-F, Boutet-Mercey S, Cachet X, De La Torre E, Laboureur L, Gallard JF, Retailleau P, Brunelle A, Dorrestein PC, Costa J, Bedoya LM, Roussi F, Leyssen P, Alcami J, Paolini J, Litaudon M, Touboul D (2017) Environmentally friendly procedure based on supercritical fluid chromatography and tandem mass spectrometry molecular networking for the discovery of potent antiviral compounds from *Euphorbia semiperfoliata*. J Nat Prod 80:2620
65. Jørgensen L, McKerrall SJ, Kuttruff CA, Ungeheuer F, Felding J, Baran PS (2015) 14-step synthesis of (+)-ingenol from (+)-3-carene (erratum report). Science 27:350
66. Bartsch H, Hecker E (1967) Circular dichroism in phorbol derivatives. Angew Chem Int Ed 6:974
67. In plants of the world online. Kew Science. https://powo.science.kew.org/taxon/urn:lsid:ipni.org:names:30375178-2

68. Roach JS, Devappa RK, Makkar HPS, Becker K (2012) Isolation, stability and bioactivity of *Jatropha curcas* phorbol esters. Fitoterapia 83:586
69. Bergquist KE, Obianwu HO, Wickberg B (1989) Isolation and structure determination of a novel phorbol derivative in an intramolecular diester macrolide. J Chem Soc Chem Commun 3:183
70. International plant name index. https://www.ipni.org/. Accessed on 14 Apr 2024
71. Zhang T, Liu Z, Sun X, Liu Z, Zhang L, Zhang Q, Peng W, Wu C (2022) Botany, traditional uses, phytochemistry, pharmacological and toxicological effects of *Croton tiglium* Linn.: a comprehensive review. J Pharm Pharmacol 74:1061
72. Taylor SE, Gafur MA, Choudhury AK, Evans FJ (1983) Sapintoxin A, a new biologically active nitrogen containing phorbol ester. Experientia 57:681
73. Brooks G, Evans AT, Aitken A, Evans FJ (1987) Sapintoxin A. A fluorescent phorbol ester that is a potent activator of protein kinase C but is not a tumour promoter. Cancer Lett 38:165
74. Ludger A, Wessjohann LA, Brandt W, Thiemann T (2003) Biosynthesis and metabolism of cyclopropane rings in natural compounds. Chem Rev 103:1625
75. Li YN, He J, Zhang J, Shi YX, Guo LB, Peng ZC, Yang T, Ding K, Zhang WK, Xu JK (2021) Existing knowledge on *Euphorbia fischeriana* Steud. (Euphorbiaceae): traditional uses, clinical applications, phytochemistry, pharmacology and toxicology. J Ethnopharmacol 275:114095
76. Sun Y, Liu JC (2011) Chemical constituents and biological activities of *Euphorbia fischeriana* STEUD. Chem Biodivers 8:1205
77. Yang DS, Peng WB, Li ZL, Wang X, Wei JG, He QX, Li XL (2014) Chemical constituents from *Euphorbia stracheyi* and their biological activities. Fitoterapia 97:211
78. Liu T, Liang Q, Zhang XM, Su XM, Qin JC, Li GQ, Xu WH (2019) Chemical constituents from *Euphorbia stracheyi* Boiss. Biochem Syst Ecol 84:52
79. Huang SZ, Zhang X, Ma QY, Zheng YT, Dai HF, Wang Q, Zhao YX (2015) Anti-HIV terpenoids from *Daphne aurantiaca* Diels. stems. RSC Adv 5:80254
80. Wang CJ, Yan QL, Ma YF, Sun CP, Chen CM, Tian XG, Ma XC (2017) *ent*-Abietane and tigliane diterpenoids from the roots of *Euphorbia fischeriana* and their inhibitory effects against *Mycobacterium smegmatis*. J Nat Prod 80:1248
81. Wang AH, Tian XG, Cui YL, Huo XK, Zhang BJ, Deng S, Wang C (2018) Diterpenoids from the roots of *Euphorbia ebracteolata* and their inhibitory effects on human carboxylesterase 2. Phytochemistry 146:82
82. Peng K, Mei WL, Zhao YX, Tan LH, Wang QH, Dai HF (2011) A novel degraded sesquiterpene from the fresh stem of *Aquilaria sinensis*. J Asian Nat Prod Res 13:951
83. Yan XL, Chen BL, Yuan FY, Zhu QF, Zhang X, Lin Y, Liao SG (2022) Six new tigliane diterpenoids with anti-inflammatory activity from *Euphorbia kansuensis*. Arab J Chem 15(6):103807
84. Fellhauer M, Hecker E (1986) Screening of Thymelaeaceae species for irritant, cocarcinogenic and antineoplastic activity. Planta Med 52:553
85. Huang SZ, Zhang X, Ma QY, Peng H, Zheng YT, Hu JM, Zhao YX (2014) Anti-HIV-1 tigliane diterpenoids from *Excoecaria acerifolia* Didr. Fitoterapia 95:34
86. Olivon F, Palenzuela H, Girard-Valenciennes E, Neyts J, Pannecouque C, Roussi F, Litaudon M (2015) Antiviral activity of flexibilane and tigliane diterpenoids from *Stillingia lineata*. J Nat Prod 78:1119
87. El-Mekkawy S, Meselhy MR, Nakamura N, Hattori M, Kawahata T, Otake T (2000) Anti-HIV-1 phorbol esters from the seeds of *Croton tiglium*. Phytochemistry 53:457
88. Van Duuren BL, Sivak A (1968) Tumor-promoting agents from *Croton tiglium* L. and their mode of action. Cancer Res 28:2349
89. Zhang XL, Wang L, Li F, Yu K, Wang MK (2013) Cytotoxic phorbol esters of *Croton tiglium*. J Nat Prod 76:858
90. Marshall GT, Klocke JA, Lin LJ, Kinghorn AD (1985) Effects of diterpene esters of tigliane, daphnane, ingenane, and lathyrane types on pink bollworm, *Pectinophora gossypiella* Saunders (Lepidoptera: Gelechiidae). J Chem Ecol 11:191

91. Fürstenberger G, Hecker E (1986) On the active principles of the Euphorbiaceae, XII. Highly unsaturated irritant diterpene esters from *Euphorbia tirucalli* originating from Madagascar. J Nat Prod 49:386
92. Zhang DD, Zhou B, Yu JH, Xu CH, Ding J, Zhang H, Yue JM (2015) Cytotoxic tigliane-type diterpenoids from *Croton tiglium*. Tetrahedron 71:9638
93. Evans FJ (1978) The irritant toxins of Blue Euphorbia (*Euphorbia coerulescens* Haw.). Toxicon 6:51
94. Rios MY, Aguilar-Guadarrama AB (2006) Nitrogen-containing phorbol esters from *Croton ciliatoglandulifer* and their effects on cyclooxygenases-1 and-2. J Nat Prod 69:887
95. Wagh VD, Korinek M, Lo IW, Hsu YM, Chen SL, Hsu HY, Chang FR (2017) Inflammation modulatory phorbol esters from the seeds of *Aquilaria malaccensis*. J Nat Prod 80:1421
96. El-Desoky AHH, Eguchi K, Kishimoto N, Asano T, Kato H, Hitora Y, Shunsuke Kotani S, Nakamura T, Tsuchiya S, Kawahara T, Watanabe M, Wada M, Nakajima M, Watanabe T, Misumi S, Tsukamoto S (2022) Isolation, synthesis, and structure–activity relationship study on daphnane and tigliane diterpenes as HIV latency-reversing agents. J Med Chem 65:3460
97. Jiang HZ, Ma QY, Huang SZ, Liang WJ, Wang PC, Zhao YX (2012) A new tigliane-type diterpene ester from *Wikstroemia scytophylla*. Chem Nat Comp 48:587
98. Asada Y, Sukemori A, Watanabe T, Malla KJ, Yoshikawa T, Li W, Lee KH (2013) Isolation, structure determination, and anti-HIV evaluation of tigliane-type diterpenes and biflavonoid from *Stellera chamaejasme*. J Nat Prod 76:852
99. Zhang M, Otsuki K, Kikuchi T, Bai ZS, Zhou D, Huang L, Li W (2021) LC-MS identification, isolation, and structural elucidation of anti-HIV tigliane diterpenoids from *Wikstroemia lamatsoensis*. J Nat Prod 84:2366
100. Liang S, Shen YH, Feng Y, Tian JM, Liu XH, Xiong Z, Zhang WD (2010) Terpenoids from *Daphne aurantiaca* and their potential anti-inflammatory activity. J Nat Prod 73:532
101. Dagang W, Sorg B, Adolf W, Opferkuch HJ, Seip EH, Hecker E (1993) Oligo-and macrocyclic diterpenes in Thymelaeaceae and Euphorbiaceae occurring and utilized in Yunnan (Southwest China), 4. Tigliane type diterpene esters (phorbol-12, 13-diesters) from *Wikstroemia canescens*. Phytother Res 7:194
102. Zhou D, Otsuki K, Zhang M, Chen G, Bai ZS, Yu H, Li N (2022) Anti-HIV tigliane-type diterpenoids from the aerial parts of *Wikstroemia lichiangensis*. J Nat Prod 85:1658
103. Hayes PY, Chow S, Somerville MJ, Fletcher MT, De Voss JJ (2010) Daphnane-and tigliane-type diterpenoid esters and orthoesters from *Pimelea elongata*. J Nat Prod 73:1907
104. Wang R, Li JY, Qi HY, Shi YP (2013) Two new tigliane diterpene esters from the flower buds of *Daphne genkwa*. J Asian Nat Prod Res 15:502
105. Hou X, Hou X, Li L (2020) Chemical constituents from the flower buds of *Daphne genkwa* (Thymelaeaceae). Biochem Syst Ecol 91:104055
106. Corlay N, Delang L, Girard-Valenciennes E, Neyts J, Clerc P, Smadja J, Litaudon M (2014) Tigliane diterpenes from *Croton mauritianus* as inhibitors of Chikungunya virus replication. Fitoterapia 97:87
107. Gunasekera SP, Kinghorn AD, Cordell GA, Farnsworth NR (1981) Plant anticancer agents. XIX. Constituents of *Aquilaria malaccensis*. J Nat Prod 44:569
108. Han BS, Van Minh N, Choi HY, Byun JS, Kim WG (2017) Daphnane and phorbol diterpenes, anti-neuroinflammatory compounds with Nurr1 activation from the roots and stems of *Daphne genkwa*. Biol Pharm Bull 40:2205
109. He Q, Zhang L, Li T, Li C, Song H, Fan P (2021) Genus *Sapium* (Euphorbiaceae): a review on traditional uses, phytochemistry, and pharmacology. J Ethnopharmacol 277:114206
110. Handa SS, Kinghorn AD, Cordell GA, Farnsworth NR (1983) Plant anticancer agents. XXII. Isolation of a phorbol diester and its $\Delta^{5,6}$-7β-hydroperoxide derivative from *Ostodes paniculata*. J Nat Prod 46:123
111. Upadhyay RR, Hecker E (1976) A new cryptic irritant and cocarcinogen from seeds of *Croton sparsiflorus*. Phytochemistry 15:1070
112. Korinek M, Wagh VD, Lo IW, Hsu YM, Hsu HY, Hwang TL, Chang FR (2016) Antiallergic phorbol ester from the seeds of *Aquilaria malaccensis*. Int J Mol Sci 17:398

113. Pei YH, Kim JW, Kang HB, Lee HK, Kim CS, Song HH, Oh SR (2012) Tigliane diterpene esters with IFN γ-inducing activity from the leaves of *Aleurites fordii*. Bioorg Med Chem Lett 22:2318
114. Hyun J, Park MH, Lee YH, Lee Y, Jeong SJ, Choi SS, Choi JH (2021) *Vernicia fordii* (Hemsl.) Airy Shaw extract stimulates insulin secretion in pancreatic β-cells and improves insulin sensitivity in diabetic mice. J Ethnopharmacol 278:114238
115. Edwards MC, Taylor SE, Williamson EM, Evans FJ (1983) New phorbol and deoxyphorbol esters: isolation and relative potencies in inducing platelet aggregation and erythema of skin. Acta Pharmacol Toxicol 53:177
116. Pan LL, Fang PL, Zhang XJ, Ni W, Li L, Yang LM, Liu HY (2011) Tigliane-type diterpenoid glycosides from *Euphorbia fischeriana*. J Nat Prod 74:1508
117. Chen H, ZhangR LRH, Yang LM, Wang RR, Hao XJ, Zheng YT (2017) Anti-HIV activities and mechanism of 12-*O*-tricosanoylphorbol-20-acetate, a novel phorbol ester from *Ostodes katharinae*. Molecules 22:1498
118. Costa LL, David VC, Pinto R, Minozzo BR, Kozlowski VA Jr, Campos LA, Beltrame FL (2012) Anti-ulcer activity of *Synadenium grantii* latex. Rev Bras Farmacogn 22:1070
119. Dewir YH, El Mahrouk ME, Hafez YM, Rihan HZ, Sáez CA, Fuller MP (2015) Antioxidative capacity and electrolyte leakage in healthy versus *Phytoplasma* infected tissues of *Euphorbia coerulescens* and *Orbea gigantea*. J Plant Physiol Pathol 3:1
120. Urones JG, Barcala PB, Cuadrado MJS, Marcos IS (1988) Diterpenes from the latex of *Euphorbia broteri*. Phytochemistry 27:207
121. Okuda T, Yoshida T, Koike S, Toh N (1975) New diterpene esters from *Aleurites fordii* fruits. Phytochemistry 14:509
122. Devkota HP, Basnet P, Yahara S (2009) Diterpene esters and phenolic compounds from *Sapium insigne* (Royle) Benth. ex Hook. Chem Pharm Bull 57:1289
123. Otsuki K, Zhang M, Yamamoto A, Tsuji M, Tejima M, Bai ZS, Zhou D, Huang L, Chen CH, Lee KH, Li N, Li W, Koike K (2020) Anti-HIV tigliane diterpenoids from *Wikstroemia scytophylla*. J Nat Prod 83:3584
124. Weber J, Hecker E (1978) Cocarcinogens of the diterpene ester type from *Croton flavens* L. and esophageal cancer in Curaçao. Experientia 34:679
125. Chen YY, Yang KX, Yang XW, Khan A, Liu L, Wang B, Zhao YL, Liu YP, Li Y, Luo XD (2016) New cytotoxic tigliane diterpenoids from *Croton caudatus*. Planta Med 82:729
126. Seip EH, Ott HH, Hecker E (1983) Skin irritant and tumor promoting diterpene esters of the tigliane type from the Chinese tallow tree (*Sapium sebiferum*). Planta Med 49:199
127. Wu XD, Zhang LC, He J, Li GT, Ding LF, Gao X, Dong LB, Song LD, Li Y, Zhao QS (2013) Two new diterpenoids from *Excoecaria acerifolia*. J Asian Nat Prod Res 15:151
128. Grosvenor SN, McLean S, Reynolds WF, Tinto WF (2007) Two phorbol esters from *Sapium hippomane*. Nat Prod Commun 2:127
129. Weng HZ, Tian Y, Zhang JS, Huang JL, Tang GH, Yin S (2021) A new tigliane-type diterpenoid from *Euphorbia tirucalli*. Nat Prod Res 36:5380
130. Fürstenberger G, Hecker E (1985) On the active principles of the spurge family (Euphorbiaceae) XI. The skin irritant and tumor promoting diterpene esters of *Euphorbia tirucalli* L. originating from South Africa. Z Naturforsch C 40:631
131. Marco JA, Sanz-Cervera JF, Checa J, Palomares E, Fraga BM (1999) Jatrophane and tigliane diterpenes from the latex of *Euphorbia obtusifolia*. Phytochemistry 52:479
132. Zhang XL, Khan AA, Wang L, Yu K, Li F, Wang MK (2016) Four new phorbol diesters from *Croton tiglium* and their cytotoxic activities. Phytochemistry Lett 16:82
133. Kinghorn AD (1980) Major skin-irritant principle from *Synadenium grantii*. J Pharm Sci 69:1446
134. Bagavathi R, Sorg B, Hecker E (1988) Tigliane-type diterpene esters from *Synadenium grantii*. Planta Med 54:506
135. Tostes JB, Carvalho AL, Ribeiro da Silva AJ, Mourão PJP, Rossi AD, Tanuri A, Siani AC (2021) Phorbol esters from the latex of *Euphorbia umbellata*: bioguided isolation of highly potent HIV-1 latency interrupters in virus reservoir cells. J Nat Prod 84:1666

136. Brooks G, Morrice NA, Ellis C, Aitken A, Evans AT, Evans FJ (1987) Toxic phorbol esters from Chinese tallow stimulate protein kinase C. Toxicon 25:1229
137. Chumkaew P, Karalai C, Ponglimanont C, Chantrapromma K (2003) Antimycobacterial activity of phorbol esters from the fruits of *Sapium indicum*. J Nat Prod 66:540
138. Taylor SE, Gafur MA, Choudhury AK, Evans FJ (1982) Sapatoxins, aliphatic ester tigliane diterpenes from *Sapium indicum*. Phytochemistry 21:405
139. Dagang W, Sorg B, Hecker E (1994) Oligo-and macrocyclic diterpenes in Thymelaeaceae and Euphorbiaceae occurring and utilized in Yunnan (Southwest China). Tigliane type diterpene esters from latex of *Euphorbia prolifera*. Phytother Res 8:95
140. Zayed SM, Farghaly M, Soliman SM, Gotta H, Sorg B, Hecker E (2001) Dietary cancer risk from conditional cancerogens (tumor promoters) in produce of livestock fed on species of spurge (Euphorbiaceae). V. Skin irriitant and tumor-promoting diterpene ester toxins of the tigliane and ingenane type in the herbs *Euphorbia nubica* and *Euphorbia helioscopia* contaminating fodder of livestock. J Cancer Res Clin Oncol 127:40
141. Falsone G, Crea AE (1979) Über Inhaltsstoffe von Euphorbiaceae, VI. Drei neue 4-Desoxyphorboltriester aus *Euphorbia biglandulosa* Desf. Liebigs Ann Chem 1979(8):1116
142. Gustafson KR, Cardellina JH II, McMahon JB, Gulakowski RJ, Ishitoya J, Szallasi Z, Lewin NE, Blumberg P, Weislow OS, Beutler JA, Buckheit RW, Cragg GM, Cox PA, Bader JB, Boyd MR (1992) A nonpromoting phorbol from the Samoan medicinal plant *Homalanthus nutans* inhibits cell killing by HIV-1. J Med Chem 35:1978
143. Adelakun TA, Ding X, Ombati RM, Zhao ND, Obodozie-Ofoegbu OO, Di YT, Zhang Y, Hao XJ (2020) A new highly oxygenated abietane diterpenoid and a new lysosome generating phorbol ester from the roots of *Euphorbia fischeriana* Steud. Nat Prod Res 34:3027
144. Wang YB, Huang R, Wang HB, Jin HZ, Lou LG, Qin GW (2006) Diterpenoids from the roots of *Euphorbia fischeriana*. J Nat Prod 69:967
145. Zayed S, Hafez A, Adolf W, Hecker E (1977) New tigliane and daphnane derivatives from *Pimelea prostrata* and *Pimelea simplex*. Experientia 33:1554
146. Sosath S, Ott HH, Hecker E (1988) Irritant principles of the spurge family (Euphorbiaceae) XIII. Oligocyclic and macrocyclic diterpene esters from latices of some *Euphorbia* species utilized as source plants of honey. J Nat Prod 51:1062
147. Baloch IB, Baloch MK, Saqib QN (2008) Anti-tumor 12-deoxyphorbol esters from *Euphorbia cornigera*. Eur J Med Chem 43:274
148. Sobottka AM, Görick C, Melzig MF (2016) Analysis of diterpenoid compounds from the latex of two Euphorbiaceae by liquid chromatography-electrospray ionisation mass spectrometry. Nat Prod Res 30:1941
149. Adolf W, Hecker E (1980) New irritant diterpene-esters from roots of *Stillingia sylvatica* L. (Euphorbiaceae). Tetrahedron Lett 21:2887
150. Adolf W, Hecker E (1982) On the active principles of the Thymelaeaceae. Planta Med 45:177
151. Hergenhahn M, Kusumoto S, Hecker E (1984) On the active principles of the spurge family (Euphorbiaceae). J Cancer Res Clin Oncol 108:98
152. Gschwendt M, Hecker E (1969) Tumor promoting compounds from *Euphorbia triangularis*: mono-and diesters of 12-desoxy-phorbol. Tetrahedron Lett 10:3509
153. Schmidt RJ, Evans FJ (1977) The succulent euphorbias of Nigeria. II. Aliphatic diterpene esters of the latices of *E. poissonii* Pax. and *E. unispina* NE Br. Lloydia 40:225
154. Evans FJ, Kinghorn AD, Schmidt RJ (1975) Some naturally occurring skin irritants. Acta Pharm Toxicol 37:250
155. Fatope MO, Zeng L, Ohayaga JE, Shi G, McLaughlin JL (1996) Selectively cytotoxic diterpenes from *Euphorbia poissonii*. J Med Chem 39:1005
156. Zhao W, Wolfender JL, Mavi S, Hostettmann K (1998) Diterpenes and sterols from *Neoboutonia melleri*. Phytochemistry 48:1173
157. Evans FJ, Schmidt RJ (1979) The succulent euphorbias of Nigeria. III. Structure and potency of the aromatic ester diterpenes of *Euphorbia poissonii* Pax. Acta Pharm Toxicol 45:181
158. Erickson KL, Beutler JA, Cardellina JH, McMahon JB, Newman DJ, Boyd MR (1995) A novel phorbol ester from *Excoecaria agallocha*. J Nat Prod 58:769

159. Abreu LS, do Nascimento YM, Costa RDS, Guedes MLS, Souza BNRF, Pena LJ, de Oliveira Costa VC, Scotti MT, Braz-Filho R, Barbosa-Filho JM, Sobral da Silva M, da Silva Velozo E, Tavares JF (2019) Tri- and diterpenoids from *Stillingia loranthacea* as inhibitors of Zika virus replication. J Nat Prod 82:2721
160. Lu Y, Huang YS, Chen CH, AkiyamaT M-N, Cheng YY, Yang SZ, Chen DF, Lee KH (2020) Anti-HIV tigliane diterpenoids from *Reutealis trisperma*. Phytochemistry 174:112360
161. Karalai C, Wiriyachitra P, Opferkuch HJ, Hecker E (1994) Cryptic and free skin irritants of the daphnane and tigliane types in latex of *Excoecaria agallocha*. Planta Med 60:351
162. Kirira PG, Rukunga GM, Wanyonyi AW, Muthaura CN, Mungai GM, Machocho AK, Ndiege IO (2007) Tigliane diterpenoids from the stem bark of *Neoboutonia macrocalyx*. J Nat Prod 70:842
163. Shi Q, Sun YW, Meng D (2017) Phytochemical and cytotoxic studies on the roots of *Euphorbia fischeriana*. Bioorg Med Chem Lett 27:266
164. Deng YY, Qu B, Zhan ZL, Wang AQ, Zhou W, Jia MY, Ha J, Luo SH (2021) Bioactive tigliane diterpenoids from the latex of *Euphorbia fischeriana*. Nat Prod Res 35:79
165. Ogura M, Koike K, Cordell GA, Farnsworth NR (1978) Potential anticancer agents VIII. Planta Med 33:128
166. Satyanarayana P, Kumar KA, Singh SK, Rao GN (2001) A new phorbol diester from *Aleurites moluccana*. Fitoterapia 72:304
167. Gschwendt M, Hecker E (1970) Tumor-promoting compounds from *Euphorbia cooperi*, di- and triesters of 16-hydroxy-12-desoxy-phorbol. Tetrahedron Lett 11:567
168. Ezzanad A, Gómez-Oliva R, Escobar-Montaño F, Díez-Salguero M, Geribaldi-Doldan N, Dominguez-Garcia S, Botubol-Ares JM, de los Reyes C, Durán-Patrón R, Nunez-Abades P, Macías-Sánchez AJ, Castro C, Hernández-Galán R (2021) Phorbol diesters and 12-deoxy-16-hydroxyphorbol 13,16-diesters induce TGFα release and adult mouse neurogenesis. J Med Chem 64:6070
169. Tietjen I, Ngwenya BN, Fotso G, Williams DE, Simonambango S, Ngadjui BT, Andersen RJ, Brockman MA, Brumme ZL, Andrae-Marobela K (2018) The *Croton megalobotrys* Müll. Arg. traditional medicine in HIV/AIDS management: documentation of patient use, in vitro activation of latent HIV-1 provirus, and isolation of active phorbol esters. J Ethnopharmacol 211:267
170. El-Sherei MM, Islam WT, El-Dine RS, El-Toumy SA, Ahmed SR (2015) Phytochemical investigation of the cytotoxic latex of *Euphorbia cooperi* NE Br. Aust J Basic Appl Sci 9:488
171. Li SF, Jiao YY, Zhang ZQ, Chao JB, Jia J, Shi XL, Zhang LW (2018) Diterpenes from buds of *Wikstroemia chamaedaphne* showing anti-hepatitis B virus activities. Phytochemistry 151:17
172. Freeman PW, Ritchie ER, Taylor W (1979) The constituents of Australian *Pimelea* spp. I. The isolation and structure of the toxin of *Pimelea simplex* and *P. trichostachya* form B responsible for St. George disease of cattle. Aust J Chem 32:2495
173. Adolf W, Hecker E (1975) On the irritant and cocarcinogenic principles of *Hippomane mancinella*. Tetrahedron Lett 16:1587
174. Chen S, Tsang NY, Jiang H, Zhang H, Ge X, Wu Y, Shen Y, Xie W, Li J (2022) Cytotoxic tigliane diterpenoids from *Euphorbia neorubella*. Nat Prod Res 36:2753
175. Appendino G, Belloro E, Tron GC, Jakupovic J, Ballero M (1999) Diterpenoids from *Euphorbia pithyusa* subsp. *cupanii*. J Nat Prod 62:1399
176. Esposito M, Nothias LF, Retailleau P, Costa J, Roussi F, Neyts J, Leyssen P, Touboul D, Litaudon M, Paolini J (2017) Isolation of premyrsinane, myrsinane, and tigliane diterpenoids from *Euphorbia pithyusa* using a Chikungunya virus cell-based assay and analogue annotation by molecular networking. J Nat Prod 80:2051
177. Karalai C, Wiriyachitra P, Sorg B, Hecker E (1995) Medicinal plants of Euphorbiaceae occurring and utilized in Thailand. V. Skin irritants of the daphnane and tigliane type in latex of *Excoecaria bicolor* and the uterotonic activity of the leaves of the tree. Phytother Res 9:482
178. Sulyok E, Vasas A, Rédei D, Dombi G, Hohmann J (2009) Isolation and structure determination of new 4,12-dideoxyphorbol esters from *Euphorbia pannonica* Host. Tetrahedron 65:4013

179. Duarte N, Lage H, Ferreir MJU (2008) Three new jatrophane polyesters and antiproliferative constituents from *Euphorbia tuckeyana*. Planta Med 74:61
180. Hegazy EF, Hamed AR, Ibrahim MA, Talat Z, Reda EH, Abdel-Azim NS, Hammouda FM, Nakamura S, Matsuda H, Haggag EG, Paré PW, Efferth T (2018) Euphosantianane A-D: antiproliferative premyrsinane diterpenoids from the endemic Egyptian plant *Euphorbia sanctae-catharinae*. Molecules 23:2221
181. Wang L, Ma YT, Sun QY, Li XN, Yan Y, Yang J, Yang FM, Liu FY, Zang Z, Wu XH, Huang SX, Zhao Y (2015) Structurally diversified diterpenoids from *Euphorbia dracunculoides*. Tetrahedron 71:5484
182. Song QQ, Rao Y, Tang GH, Sun ZH, Zhang JS, Huang ZS, Yin S (2019) Tigliane diterpenoids as a new type of antiadipogenic agents inhibit GRα-Dexras1 axis in adipocytes. J Med Chem 62:2060
183. Wang HB, Chu WJ, Wang Y, Ji P, Wang YB, Yu Q, Qin GW (2010) Diterpenoids from the roots of *Euphorbia fischeriana*. J Asian Nat Prod Res 12:1038
184. Hassan EM, Mohammed MD, Mohamed SM (2012) Two new phorbol-type diterpene esters from *Synadenium grantii* Hook. f. leaves. Rec Nat Prod 3:255
185. Campos A, Vendramini-Costa DB, Longato GB, Zermiani T, Ruiz ALTG, de Carvalho JE, Cechinel Filho V (2016) Antiproliferative effect of *Synadenium grantii* Hook f. stems (Euphorbiaceae) and a rare phorbol diterpene ester. Int J Toxicol 35:666
186. Dai LF, Liang Q, Liu T, He MY, Zhao P, Xu WH (2016) A new tigliane-type diterpene from *Euphorbia dracunculoides* Lam. Nat Prod Res 30:1639
187. Cui JJ, Ji KL, Liu HC, Zhou B, Liu QF, Xu CH, Ding J, Zhao JX, Yue JM (2019) Cytotoxic tigliane diterpenoids from *Croton damayeshu*. J Nat Prod 82:1550
188. Zolfaghari B, Yazdiniapour Z, Ghanadian M, Lanzotti V (2016) Cyclomyrsinane and premyrsinane diterpenes from *Euphorbia sogdiana* Popov. Tetrahedron 72:5394
189. Ghanadian M, Ayatollahi AM, Mesaik MA, Afsharypuor S, Abdalla OM, Kobarfard F (2011) New tigliane-type diterpenoids from *Euphorbia aellenii* Rech. with immunomodulatory activity. Res Pharm Sci 6:35
190. Yazdiniapour Z, Ghanadian M, Zolfaghari B, Lanzotti V (2016) 6(17)-Epoxylathyrane diterpenes from *Euphorbia sogdiana* Popov with cytotoxic activity. Fitoterapia 108:87
191. Zhao BQ, Peng S, He WJ, Liu YH, Wang JF, Zhou XJ (2016) Antitubercular and cytotoxic tigliane-type diterpenoids from *Croton tiglium*. Bioorg Med Chem Lett 26:4996
192. Pettit GR, Ducki S, Tan R, Gardella RS, McMahon JB, Boyd MR, Pettit GR III, Blumberg PM, Lewin NE, Doubek DL, Tackett LP, Williams MD (2002) Isolation and structure of pedilstatin from a Republic of Maldives *Pedilanthus* sp. J Nat Prod 65:1262
193. Ndunda B, Langat MK, Wanjohi JM, Midiwo JO, Kerubo LO (2013) Alienusolin, a new 4α-deoxyphorbol ester derivative, and crotonimide C, a new glutarimide alkaloid from the Kenyan *Croton alienus*. Planta Med 79:1762
194. Haba H, Lavaud C, Harkat H, Magid AA, Marcourt L, Benkhaled M (2007) Diterpenoids and triterpenoids from *Euphorbia guyoniana*. Phytochemistry 68:1255
195. Aljancic IS, Pešić M, Milosavljevic SM, Todorovic NM, Jadranin M, Milosavljevic G, Povrenović D, Banković J, Tanić N, Marković ID, Ruzdijić S, Vajs VE, Tešević VV (2011) Isolation and biological evaluation of jatrophane diterpenoids from *Euphorbia dendroides*. J Nat Prod 74:1613
196. Chen R, You CX, Wang Y, Zhang WJ, Yang K, Geng ZF, Liu ZL, Deng ZW, Wang YY, Du SS (2014) Chemical constituents from the roots of *Euphorbia nematocypha* Hand.-Mazz. Biochem Syst Ecol 57:1
197. Benmerache A, Magid A, Labed A, Kabouche A, Voutquenne-Nazabadioko L, Morjani HJ, Kabouche Z (2017) Isolation and characterisation of cytotoxic compounds from *Euphorbia clementei* Boiss. Nat Prod Res 31:2091
198. Aichour S, Haba H, Benkhaled M, Harakat D, Lavaud C (2014) Terpenoids and other constituents from *Euphorbia bupleuroides*. Phytochem Lett 10:198
199. Ren FX, Ren FZ, Yang Y, Yu NJ, Zhang Y, Zhao YM (2014) Tigliane diterpene esters from the leaves of *Croton tiglium*. Helv Chim Acta 97:1014

200. Forgo P, Rédei D, Hajdu Z, Szabó P, Szabó L, Hohmann J (2011) Unusual tigliane diterpenes from *Euphorbia grandicornis*. J Nat Prod 74:639
201. Wang J, Qin L, Zhao B, Cai L, Zhong Z, Liu Y, Zhou X (2019) Crotonols A and B, two rare tigliane diterpenoid derivatives against K562 cells from *Croton tiglium*. Org Biomol Chem 17:195
202. Jiang ZY, Feng JE, Duan LK, Liu CJ, Li XF, Huang CQ, Shi SL, Wan RR, Zuo AX, He HP (2022) Tigliane diterpenoids with larvicidal, antifungal, and α-glucosidase inhibitory activities from *Croton damayeshu*. J Nat Prod 85:405
203. Ma QY, Chen YC, Huang SZ, Kong FD, Zhou LM, Dai HF, Hua Y, Zhao YX (2016) Chemical constituents from the stems of *Daphne holosericea* (Diels) Hamaya. Chem Biodivers 13:1469
204. Li DY, Lee C, Jin Q, Lee JW, Lee MK, Hwang BY (2014) A new tigliane-type diterpenoid from *Daphne genkwa*. Bull Kor Chem Soc 35:669
205. Bourjot M, Delang L, Nguyen VH, Neyts J, Guéritte F, Leyssen P, Litaudon M (2012) Prostratin and 12-*O*-tetradecanoylphorbol 13-acetate are potent and selective inhibitors of Chikungunya virus replication. J Nat Prod 75:2183
206. Liang X, Liu ZG, Cao YF, Meng DL, Hua HM (2014) Chemotaxonomic and chemical studies on two plants from genus of *Euphorbia*: *Euphorbia fischeriana* and *Euphorbia ebracteolata*. Biochem Syst Ecol 57:345
207. Ferreira RJ, Spengler G, Orthaber A, dos Santos AJVA, Ferreira MJU (2021) Pedrolane, a polycyclic diterpene scaffold containing a bicyclo[2.2.1]heptane system, from *Euphorbia pedroi*. Org Lett 23:274
208. Wang L, Yang J, Kong LM, Deng J, Xiong Z, Huang J, Luo J, Yan Y, Hu Y, Li XN, Li Y, Zhao Y, Huang SX (2017) Natural and semisynthetic tigliane diterpenoids with new carbon skeletons from *Euphorbia dracunculoides* as a Wnt signaling pathway inhibitor. Org Lett 19:3911
209. Carrier M (2006) Kant's theory of matter and his views on chemistry. In: Watkins E (ed) Kant and the sciences. Oxford Academic, New York, p 205
210. Luo D, Callari R, Hamberger B, Wubshet SG, Nielsen MT, Andersen-Ranberg J, Hallström BM, Cozzi F, Heider H, Møller BL, Staerk D, Hamberger B (2016) Oxidation and cyclization of casbene in the biosynthesis of *Euphorbia* factors from mature seeds of *Euphorbia lathyris* L. Proc Natl Acad Sci USA 113:5082
211. Mitu SA, Ogbourne SM, Klein AH, Tran TD, Reddell PW, Cummins SF (2021) The P450 multigene family of *Fontainea* and insights into diterpenoid synthesis. BMC Plant Biol 21:191
212. Wong J, de Rond T, d'Espaux L, van der Horst C, Dev I, Rios-Solis L, Kirby J, Scheller H, Keasling J (2018) High-titer production of lathyrane diterpenoids from sugar by engineered *Saccharomyces cerevisiae*. Metab Eng 45:142
213. Adolf W, Hecker E (1975) On the active principles of the spurge family. III. Skin irritant and cocarcinogenic factors from the caper spurge. Z Krebsforsch Klin Onkol Cancer Res Clin Oncol 84:325
214. Vela F, Ezzanad A, Hunter AC, Macías-Sánchez AJ, Hernández-Galán R (2022) Pharmacological potential of lathyrane-type diterpenoids from phytochemical sources. Pharmaceuticals 15:780
215. Wu SQ, Zhu X, Yuan T, Yuan FY, Zhou S, Huang D, Wang Y, Tang GH, Huang SZ, Chen X, Yin S (2023) Discovery of ingenane diterpenoids from *Euphorbia hylonoma* as antiadipogenic agents. J Nat Prod 86:2691
216. Appendino G, Tron GC, Cravotto G, Palmisano G, Annunziata R, Baj G, Surico N (1999) Synthesis of modified ingenol esters. Eur J Org Chem 1999:3413
217. Mendes E, Cátia Ramalhete C (2024) Myrsinane-type diterpenes: a comprehensive review on structural diversity, chemistry, and biological activities. Int J Mol Sci 25:147
218. Wu C, Zhang J, Liu M, Xie X, Li H, She X (2023) Rearrangement of the tetra- and tricyclic skeletons of pepluanol B to access the core structures of tigliane- and myrsinane-type *Euphorbia* diterpenes. Org Lett 25:7995
219. Jensen K, Moller BL (2010) Plant NAPH-cytochrome P450 oxidoreductases. Rebound oxygenation mechanism. Phytochemistry 71:132

220. Talele TT (2016) The "cyclopropyl" fragment is a versatile player that frequently appears in preclinical/clinical drug molecules. J Med Chem 59:8712
221. Walborsky HM (1981) The cyclopropyl radical. Tetrahedron 37:1625
222. Genossar N, Changala PB, Gans B, Loison JC, Hartweg S, Martin-Drumel MA, Garcia GA, Stanton JF, Ruscic B, Baraban JH (2022) Ring-opening dynamics of the cyclopropyl radical and cation: the transition state nature of the cyclopropyl cation. J Am Chem Soc 144:18518
223. Wessjohann LA, Brandt W, Thiemann T (2003) Biosynthesis and metabolism of cyclopropane rings in natural compounds. Chem Rev 103:1625
224. Ma S, Mandalapu D, Wang S, Zhang Q (2022) Biosynthesis of cyclopropane in natural products. Nat Prod Rep 39:926
225. Van de Poel B, van Der Straeten D (2014) 1-Aminocyclopropane-1-carboxylic acid (ACC) in plants: more than just the precursor of ethylene! Front Plant Sci 5:640
226. Houben M, Van de Poel B (2019) 1-Aminocyclopropane-1-carboxylic acid oxidase (ACO): the enzyme that makes the plant hormone ethylene. Front Plant Sci 10:695
227. Tang JW, Liu X, Ye W, Li ZR, Qian PY (2022) Biosynthesis and bioactivities of microbial genotoxin colibactins. Nat Prod Rep 39:991
228. Ullrich M, Guenzi AC, Mitchell RE, Bender CL (1994) Cloning and expression of genes required for coronamic acid (2-ethyl-1-aminocyclopropane 1-carboxylic acid), an intermediate in the biosynthesis of the phytotoxin coronatine. Appl Environ Microbiol 60:2890
229. Mitchell RE, Pirrung MC, McGeehan GM (1987) Absolute configuration of norcoronatine. Phytochemistry 26:2695
230. Shiraishi K, Ichihara A, Sakamura S (1977) Facile stereoselective synthesis of (+)- and (–)-allocoronamic acids. Agric Biol Chem 41:2497
231. Winn M, Rowlinson M, Wang F, Bering L, Francis D, Levy C, Micklefield J (2021) Discovery, characterization and engineering of ligases for amide synthesis. Nature 539:391
232. Höfer I, Crüsemann M, Radzom M, Geers B, Flachshaar D, Cai X, Zeeck A, Piel J (2011) Insights into the biosynthesis of hormaomycin, an exceptionally complex bacterial signalling metabolite. Chem Biol 18:381
233. Engelbrecht A, Wolf F, Esch A, Kulik A, Kozhushkov SI, de Meijere A, Hughes HC, Kaysser L (2022) Discovery of a cryptic nitro intermediate in the biosynthesis of the 3-(*trans*-2′-aminocyclopropyl)alanine moiety of belactosin A. Org Lett 24:736
234. Kainz KP, Virtbauer J, Khlig H, Arion V, Donath O, Reznicek G, Huber W, Mariane B, Krenn L (2012) Two unusual methylidenecyclopropane glucosides from *Metaxya rostrata* C Presl. Helv Chim Acta 95:1531
235. Severino VGP, Braga PAC, da Silva MF, Fernandes JB, Vieira PC, Theodoro JE, Ellena JA (2012) Cyclopropane- and spirolimonoids and related compounds from *Hortia oreadica.* Phytochemistry 76:52
236. Shen W, Li L, Liu QH, Cui JM, Shi W, Shi XH, Zhang XQ, Ye WC, Hu XL, Wang H (2024) Characteristic chromanone acids from *Calophyllum membranaceum*: determination of C-3 configuration and anti-inflammatory activity. Phytochemistry 217:113902
237. Trottmann F, Ishida K, Ishida-Ito M, Kries H, Groll M, Hertweck C (2022) Pathogenic bacteria remodel central metabolic enzyme to build a cyclopropanol warhead. Nat Chem 14:884
238. Ueoka R, Bortfeld-Miller M, Morinaka BI, Vorholt JA, Piel J (2018) Toblerols: cyclopropanol-containing polyketide modulators of antibiosis in methylobacteria. Angew Chem Int Ed 57:977
239. Umezawa H, Muraoka Y, Fujii A, Naganawa H, Takita T (1980) Chemistry of bleomycin. XXVII. Cleomycin, a new family of bleomycin-phleomycin group. J Antibiot 33:1079
240. Hashizume H, Iijima K, Yamashita K, Kimura T, Wada S, Sawa R, Igarashi M (2018) Valgamicin C, a novel cyclic depsipeptide containing the unusual amino acid cleonine, and related valgamicins A, T and V produced by *Amycolatopsis* sp. ML1-hF4. J Antibiot 71:129
241. Snyman LD, Kellerman TS, Vleggaar R, Flett BC, Basson KM, Schultz RA (2011) Diplonine, a neurotoxin isolated from cultures of the fungus *Stenocarpella maydis* (Berk.) Sacc. that induces diplodiosis. J Agric Food Chem 59:9039

242. Aim A, Nair BG, Rabenstein J, Davis G, Baker DD (2000) CD45 Tyrosine phosphatase inhibitory components from *Aspergillus niger*. J Antibiot 53:110
243. Li X, Awakawa T, Mori T, Ling M, Hu D, Wu B, Abe I (2021) Heterodimeric non-heme iron enzymes in fungal meroterpenoid biosynthesis. J Am Chem Soc 143:21425
244. Lindberg P, Bergman R, Wickberg B (1975) Isolation and structure of coprine, a novel physiologically active cyclopropanone derivative from *Coprinus atramentarius* and its synthesis via 1-aminocyclopropanol. J Chem Soc Chem Commun 23:946
245. Appendino G (2016) Ingenane diterpenoids. Prog Chem Org Nat Prod 102:1
246. Kulinkovich OG (2003) The chemistry of cyclopropanols. Chem Rev 103:2597
247. Liu L, Montgomery J (2006) Dimerization of cyclopropyl ketones and crossed reactions of cyclopropyl ketones with enones as an entry to five-membered rings. J Am Chem Soc 128:5348
248. Asfaw N, Storesund H, Skattebøl L, Aasen A J (2001) Coexistence of chrysanthenone, filifolone and (*Z*)-isogeranic acid in hydrodistillates. Artefacts! Phytochemistry 58:489
249. Dauben WG, Friedrich LE, Oberhänsli P, Aoyagi EI (1972) Thujopsene rearrangements. The cyclopropylcarbinyl system. J Org Chem 37:9
250. Dauben WG, Thiessen WE, Resnick PR (1962) Cembrene, a 14-membered ring diterpene hydrocarbon. J Am Chem Soc 84:2015
251. Meguro A, Motoyoshi Y, Teramoto K, Ueda S, Totsuka Y, Ando Y, Tomita T, Kim SY, Kimura T, Igarashi M, Sawa R, Shinada T, Nishiyama M, Kuzuyama T (2015) An unusual terpene cyclization mechanism involving a carbon–carbon bond rearrangement. Angew Chem Int Ed 54:4353
252. Kim SY, Zhao P, Igarashi M, Sawa R, Tomita T, Nishiyama M, Kuzuyama T (2009) Cloning and heterologous expression of the cyclooctatin biosynthetic gene cluster afford a diterpene cyclase and two P450 hydroxylases. Chem Biol 16:736
253. Sato H, Li BX, Takagi T, Want C, Miyamoto K, Uchiyama M (2021) DFT study on the biosynthesis of verrucosane diterpenoids and mangicol sesterterpenoids: involvement of secondary-carbocation-free reaction cascades. JACS Au 1:123135
254. Zuo Q, Mu HY, Gong Q, Ding X, Wang W, Zhang HY, Zhao WM (2021) Diterpenoids from the seeds of *Euphorbia lathyris* and their effects on microglial nitric oxide production. Fitoterapia 150:104834
255. Hua W, Hu H, Chen F, Tang L, Peng T, Wang Z (2015) Rapid isolation and purification of phorbol esters from *Jatropha curcas* by high-speed countercurrent chromatography. J Agric Food Chem 63:2767
256. Appendino G, Jakupovic J, Varese M, Bombardelli E (1996) Acid and base catalysed rearrangements of 9,10-dioxotaxanes. Tetrahedron Lett 37:727
257. Wilson SR, Huffman JC (1976) The structural relationship of phorbol and cortisol, a possible mechanism for the tumor-promoting activity of phorbol. Experientia 32:1489
258. Hecker E, Härle E, Schairer HU, Jacobi P, Hoppe W, Gassmann J, Röhrl M, Abel H (1968) Lumiphorboltriacetat, ein käfigartiges Derivat des Diterpens 4α-Phorbols. Angew Chem Int Ed Engl 7:890
259. Robles O, Serna-Saldívar SO, Gutiérrez-Uribe JA, Romo D (2012) Cyclopropanation of olefin-containing natural products for simultaneous arming and structure activity studies. Org Lett 14:1394
260. Chow S, Krainz T, Bettencourt CJ, Broit N, Ferguson B, Zhu M, Hull KG, Pierens GK, Bernhardt PV, Parsons PG, Romo D, Boyle GM, Williams CM (2020) Synthetic tigliane intermediates engage thiols to induce potent cell line selective anti-cancer activity. Chem Eur J 26:13372
261. Schairer HU, Thielmann HW, Gschwendt M, Kreibich G, Schmidt R, Hecker E (1968) Zur Chemie des Phorbols, IV. Polybenzoate und -acetate des Phorbols und Phorbol-3-ols und funktionelle Derivate der Allylgruppierung des Phorbols. Z Naturforsch 23b:1430
262. Wender PA, Irie K, Miller BL (1993) Synthesis and binding of photoaffinity ligand candidates for protein-kinase-C. J Org Chem 58:4179

263. Appendino G, Bertolino A, Minassi A, Annunziata A, Szallasi A, de Petrocellis L, Di Marzo V (2004) Synthesis and biological evaluation of phorbol-resiniferatoxin (RTX) hybrids. Eur J Org Chem 2004(16):3413
264. Gschwendt M, Härle E, Hecker E (1968) Zur Chemie des Phorbols, VII. Die α,β-ungesättigte tertiäre 1,2-Ketolgruppierung im Phorbol. Z Naturforsch 23b:1579
265. Jacobi P, Härle E, Schairer HU, Hecker E (1970) Zur Chemie des Phorbols, XVI. 4*α*-Phorbol. Liebigs Ann Chem 741:13
266. Fayet A, Bourcier S, Casaretto H, Nay B, Frison G (2023) Theoretical investigation of the mechanism of the Hock rearrangement with $InCl_3$ as catalyst. ChemCatChem 15:e2023005
267. Hock H, Lang S (1944) Über Peroxyde von Benzol-Derivaten. Chem Ber 77:257
268. Kuroda C, Eriko Nishio E (2007) Substituent effect in color of Ehrlich's test of tetrahydrobenzofuran. Nat Prod Commun 2:581
269. Kreibich G, Hecker E (1968) Zur Chemie des Phorbols, V. Über einige Äther des Phorbols. Z Naturforsch 23b:1444
270. He W, Cik M, Appendino G, Puyvelde LV, Leysen JE, De Kimpe N (2002) Daphnane-type diterpene orthoesters and their biological activities. Mini Rev Med Chem 2:185
271. Schmidt R, Hecker E (1975) Autooxidation of phorbol esters under normal storage conditions. Cancer Res 35:1375
272. Kongmany S, Feruta M, Matsura H, Okuda S, Imamura K, Maeda Y (2014) Degradation of phorbol 12,13-diacetate in aqueous solution by gamma radiation. Radiat Phys Chem 105:98
273. Katti SS, Krieger IV, Ann J, Lee J, Sacchettini JC, Igumenova TI (2022) Structural anatomy of Protein Kinase C C1 domain interactions with diacylglycerol and other agonists. Nat Commun 13:2695
274. Kauffmann T, Jasching W, Schulz J (1966) Zur Konstitution des Phorbols, III. Epoxydierung von Phorbol und Phorbol-Derivaten. Chem Ber 99:1569
275. Tyler MI, Howden MEH (1987) Synthesis and structure-activity studies of potential antitumour agents derived from esters of phorbol. Aust J Chem 40:193
276. Kennedy AR (1995) Effects of the *6β-7β*-epoxide of 12-*O*-tetradecanoyl-phorbol-13-acetate on radiation transformation in vitro. Cancer Lett 95:7
277. Wender PA, Buschmann N, Cardin NB, Jones LR, Kan C, Kee JM, Kowalski JA, Longcore KE (2012) A gateway synthesis, PKC affinities and cell growth activities of daphnane congeners. Nat Chem 3:615
278. Boudreault PL, Mattler JK, Wender PA (2015) Studies on the regio- and diastereo-selective epoxidation of daphnanes and tiglianes. Tetrahedron Lett 56:3423
279. Arifoglu M, Marmer WN, Dudley RL (1992) Reaction of thiourea with hydrogen peroxide: ^{13}C NMR studies of an oxidative /reductive bleaching process. Textile Res J 62:94
280. Ferrier RJ, Overend WG, Ryan AE (1962) The reaction between 3,4,6-tri-*O*-acetyl-D-glucals and *p*-nitrophenol. J Chem Soc 3667
281. Morril C, Beutner GL, Grubbs RH (2006) Rhenium-catalyzed 1,3-isomerization of allylic alcohols: scope and chirality transfer. J Org Chem 71:7813
282. Amin HIM, Maioli C, Chianese G, Appendino G, Gaeta S, Taglialatela-Scafati O (2022) The allylic oxidation of tigliane esters. Fitoterapia 148:104802
283. Thielmann HW, Hecker E (1970) Zur Chemie des Phorbols XV. Oxydation der Δ^6 Doppelbindung des Phorbols mit Osmiumtetroxid und Spaltung einiger 6*β*,7*β*-Diole. Liebigs Ann Chem 735:113
284. Hecker E, von Szczepanski C, Kubinyi H, Bresch H, Härle E, Schairer HU, Bartsch H (1966) Über die Wirkstoffe des Crotonöls VII. Phorbol. Z Naturforsch 21b:1204
285. Kreibich G, Hecker E (1970) Active principles of Croton oil. X. Preparation of tritium-labelled croton oil factor A1 and other tritium labeled phorbol derivatives. Z Krebsforsch 74:448
286. Hecker E, Bartsch H, Kreibich G, von Szczepanski V (1967) Zur Chemie des Phorbols, X. Schweratom-Derivate zur Röntgenstruckturanalyse des Phorbols. Tetrahedron Lett 3165
287. Schmidt R, Hecker E (1971) Untersuchungen über die Beziehungen zwischen Struktur und Wirkung von Phorbolestern. In: Lettré H, Wagner G (eds) Aktuelle Probleme auf dem Gebiet der Cancerologie, vol III. Springer, Berlin, Heidelberg, New York, p 98

288. Paal C, Roth K (1909) Über katalytische Wirkungen kolloider Metalle der Platingruppe. V. Die Reduktion der Fette. Ber Dtsch Chem Ges 42:1541
289. Thielmann HW, Hecker E (1969) Zur Chemie des Phorbols XIV. Die Flaschenträger-Reaktion. Liebigs Ann Chem 728:158
290. Smith PAS, Baer DR (1960) The Demjanov and Tiffeneau-Demjanov ring expansions. Org React 11:157
291. Bartsch H, Hecker E (1969) Zur Chemie des Phorbols, XI. Crotophorbolon-enolacetate und Acetoxycrotophorbolon. Z Naturforsch 24b:91
292. Abdel-Hafez AA, Nakamura N, Hattori M (2002) Biotransformation of phorbol by human intestinal bacteria. Chem Pharm Bull 50:160
293. Yu T, Sun Y, Tu C, Chen T, Fu S, Liu B (2020) Total synthesis of crotophorbolone. Chem Sci 11:7177
294. Asaba T, Katoh Y, Urabe D, Inoue M (2015) Total synthesis of crotophorbolone. Angew Chem Int Ed 54:14457
295. Wender PA, Kee JM, Warrington JM (2008) Practical synthesis of prostratin, DPP, and their analogs, adjuvant leads against latent HIV. Science 320:649
296. Irie K, Nakahara A, Ikawa Y, Tanaka M, Nakagawa Y, Nakamura Y, Ohigashi H, Wender PA (2000) Synthesis and tumor-promoting activities of 12-*epi*-phorbol-12,13-dibutyrate. Biosci Biotechnol Biochem 64:2429
297. Wender PA, Kogen H, Lee HY, Munger JD, Wilhelm RS, Williams PD (1989) Studies on tumor promoters. 8. The synthesis of phorbol. J Am Chem Soc 111:8957
298. Bartsch H, Hecker E (1969) Zur Chemie des Phorbols XIII. Über eine Acyloin-Umlagerung des 12-Desoxy-12-oxophorbol-13,20-diacetats. Liebigs Ann Chem 725:142
299. Alabugin IV (2016) Stereoelectronic effects. A bridge between structure and reactivity. Wiley VCH, Hoboken, NJ
300. Neumann H (1959) Zur Konstitution des Phorbols. I. Über die reduzierende Gruppe des Phorbols. Chem Ber 92:1715
301. Yan H, Smith GS, Chen FE (2022) Recent advances using cyclopropanols and cyclobutanols in ring-opening asymmetric synthesis. Green Synth Cat 3:2019
302. Gschwendt M, Hecker E (1968) Zur Chemie des Phorbols, IX. Oxydation von Phorbol mit Bleitetraacetat. Z Naturforsch 23b:1584
303. Gschwendt M, Hecker E (1969) Zur Chemie des Phorbols, VIII. Oxydation von Phorbol mit Natriumperjodat. Z Naturforsch 24b:80
304. Szczepanski Ch, Schairer HU, Gschwendt M, Hecker E (1967) Zur Chemie des Phorbols, III. Mono-und Diacetate des Phorbols. Liebigs Ann Chem 705:199
305. Casiraghi G, Zanardi F, Appendino G, Rassu G (2000) The vinylogous aldol reaction: a valuable, yet understated carbon-carbon bond-forming reaction. Chem Rev 100:1929
306. Maioli C, Hawraz IMA, Chianese G, Minassi A, Reddell PW, Gaeta S, Taglialatela-Scafati O, Appendino G (2023) Novel skeletal rearrangements of the tigliane diterpenoid core. J Nat Prod 18:2685
307. Kawamura M, Kamo S, Azuma S, Kubo K, Sasamori T, Tokitoh N, Kuramochi K, Tsubaki K (2017) Skeletal rearrangement of polycyclic α-ketols. Org Lett 19:303
308. Vasas A, Hohmann J (2014) *Euphorbia* diterpenes: isolation, structure, biological activity, and synthesis (2008–2012). Chem Rev 114:8579
309. Wang HB, Wang XY, Liu LP, Qin GW, Kang TG (2015) Tigliane diterpenoids from the Euphorbiaceae and Thymelaeaceae families. Chem Rev 115:2975
310. Liu Z, Ding Z, Chen K, Xu M, Yu T, Tong G, Zhang H, Li P (2021) Balancing skeleton and functional groups in total syntheses of complex natural products: a case study of tigliane, daphnane and ingenane diterpenoids. Nat Prod Rep 38:1589
311. Xu Y, Tang P, Zhu M, Wang Y, Sun D, Li H, Chen L (2021) Diterpenoids from the genus *Euphorbia*: structure and biological activity (2013–2019). Phytochemistry 190:112846
312. Kemboi D, Siwe-Noundou X, Krause RWM, Langat MK, Tembu VJ (2021) *Euphorbia* diterpenes: an update of isolation, structure, pharmacological activities and structure-activity relationship. Molecules 26:5055

313. Zhan ZJ, Li S, Chu W, Yin S (2022) *Euphorbia* diterpenoids: isolation, structure, bioactivity, biosynthesis, and synthesis (2013–2021). Nat Prod Rep 39:2132
314. Peplov M (2017) What counts as a step during chemical synthesis? Chem World. https://www.chemistryworld.com/opinion/what-counts-as-a-step-during-chemical-synthesis/2500261.article
315. Appendino G, Minassi O, Taglialatela-Scafati O (2014) Recreational drug discovery: natural products as lead structures for the synthesis of smart drugs. Nat Prod Rep 31:880
316. Wender PA, McDonald FE (1990) Studies on tumor promoters. 9. A second-generation synthesis of phorbol. J Am Chem Soc 112:4956
317. Wender PA, Rice KD, Schnute ME (1997) The first formal asymmetric synthesis of phorbol. J Am Chem Soc 119:7897
318. Wender PA, Lee HY, Wilhelm RS, Williams PD (1989) Studies on tumor promoters. The synthesis of a potentially general precursor of the tiglianes, daphnanes, and ingenanes. J Am Chem Soc 111:8954
319. Trost BM, Rise F (1987) A reductive cyclization of 1,6- and 1,7-enynes. J Am Chem Soc 109:3161
320. Negishi E, Holmes SJ, Tour JM, Miller JA, Cederbaum FE, Swanson DR, Takahashi T (1989) Novel bicyclization of enynes and diynes promoted by zirconocene derivatives and conversion of zirconabicycles into bicyclic enones via carbonylation. J Am Chem Soc 111:3336
321. Lee K, Cha JK (2001) Formal synthesis of (+)-phorbol. J Am Chem Soc 123:5590
322. Kawamura S, Chu H, Felding J, Baran PS (2016) Nineteen-step total synthesis of (+)-phorbol. Nature 532:90
323. Jørgensen L, McKerrall SJ, Kuttruff CA, Ungeheuer F, Felding J, Baran PS (2013) 14-Step synthesis of (+)-ingenol from (+)-3-carene. Science 341:878
324. Isayama S, Mukaiyama T (1989) A new method for preparation of alcohols from olefins with molecular oxygen and phenylsilane by the use of bis(acetylacetonato)cobalt(II). Chem Lett 18:1071
325. Inoki S, Kato K, Isayama S, Mukaiyama T (1990) A new and facile method for the direct preparation of α-hydroxycarboxylic acid esters from α, β-unsaturated carboxylic acid esters with molecular oxygen and phenylsilane catalyzed by bis(dipivaloylmethanato)manganese(II) complex. Chem Lett 19:1869
326. Tokuyasu T, Kunikawa S, Masuyama A, Nojima M (2002) Co(III)−alkyl complex- and Co(III)−alkylperoxo complex-catalyzed triethylsilylperoxidation of alkenes with molecular oxygen and triethylsilane. Org Lett 4:3595
327. Newhouse T, Baran PS (2011) If C-H bonds could talk: selective C-H bond oxidation. Angew Chem Int Ed 50:3362
328. Michaudel Q, Journot G, Regueiro-Ren A, Goswami A, Guo Z, Tully TP, Zou L, Ramabhadran RO, Houk KN, Baran PS (2014) Improving physical properties via C-H oxidation: chemical and enzymatic approaches. Angew Chem Int Ed 53:12091
329. Zou L, Paton RS, Eschenmoser A, Newhouse TR, Baran PS, Houk KN (2013) Enhanced reactivity in dioxirane C-H oxidations via strain release: a computational and experimental study. J Org Chem 78:4037
330. Sha CK, Huang SJ (1995) Synthesis of β-substituted α-iodocycloalkenones. Tetrahedron Lett 36:6927
331. Stille JK (1986) The palladium-catalyzed cross-coupling reactions of organotin reagents with organic electrophiles. Angew Chem Int Ed Eng 25:508
332. Tong G, Liu Z, Li P (2018) Total synthesis of (±)-prostratin. Chem 4:2944
333. Tong G, Ding Z, Liu Z, Ding YS, Xu L, Zhang H, Li P (2020) Total synthesis of prostratin, a bioactive tigliane diterpenoid: access to multi-stereocenter cyclohexanes from a phenol. J Org Chem 85:4813
334. Hirose A, Watanabe A, Ogino K, Nagatomo M, Inoue M (2021) Unified total syntheses of rhamnofolane, tigliane, and daphnane diterpenoids. J Am Chem Soc 143:12387
335. Nicolaou KC, Yang Z, Liu JJ, Ueno H, Nantermet PG, Guy RK, Claiborne CF, Renaud J, Couladouros EA, Paulvannan K, Sorensen EJ (1994) Total synthesis of taxol. Nature 367:630

336. Chow S, Krainz T, Bernhardt PV, Williams CM (2019) En route to D-ring inverted phorbol esters. Org Lett 21:8761
337. Kazanietz MG (2002) Novel "nonkinase" phorbol ester receptors: the C1 domain connection. Mol Pharmacol 61:759
338. Pagani A, Watanabe H, Davis JB, Smart D, Jerman JC, Smith GD, Hayes P, Vriens J, Cairns W, Wissenbach U, Prenen J, Flockerzi V, Droogmans G, Benham CD, Nilius B (2002) Activation of TRPV4 channels (hVRL-2/mTRP12) by phorbol derivatives. J Biol Chem 277:13569
339. EFSA Panel on Contaminants in the Food Chain (CONTAM) (2015) Risks for human and animal health related to the presence of phorbol esters in *Jatropha* kernel meal. EFSA J 13:4321
340. Wang XH, Liu JQ, Chen S, Yin Y, Liu Y, Zhang C (2020) Hydroxy-octadecenoic acids instead of phorbol esters are responsible for the *Jatropha curcas* kernel cake's toxicity. Commun Biol 8:228
341. Takai Y, Kishimoto A, Kikkawa U, Mori T, Nishizuka Y (1979) Unsaturated diacylglycerol as a possible messenger for the activation of calcium-activated, phospholipid-dependent protein kinase system. Biochem Biophys Res Commun 91:1218
342. Castagna M, Takai Y, Kaibuchi K, Sano K, Kikkawa U, Nishizuka Y (1982) Direct activation of calcium-activated, phospholipid-dependent protein kinase by tumor-promoting phorbol esters. J Biol Chem 257:7847
343. Newton AC (2001) Protein kinase C: structural and spatial regulation by phosphorylation, cofactors, and macromolecular interactions. Chem Rev 101:2353
344. Ikin AF, Causevic M, Pedrini S, Benson LS, Buxbaum JD, Suzuki T, Lovestone S, Higashiyama S, Mustelin T, Burgoyne RD, Gandy S (2007) Evidence against roles for phorbol binding protein Munc13-1, ADAM adaptor Eve-1, or vesicle trafficking phosphoproteins Munc18 or NSF as phospho-state-sensitive modulators of phorbol/PKC-activated Alzheimer APP ectodomain shedding. Mol Neurodegener 2:23
345. Geiger M, Wrulich OA, Jenny M, Schwaiger W, Grunicke HH, Uberall F (2003) Defining the human targets of phorbol ester and diacylglycerol. Curr Opin Mol Ther 5:631
346. Cooke M, Magimaidas A, Casado-Medrano V, Kazanietz MG (2017) Protein kinase C in cancer: the top five unanswered questions. Mol Carcinog 56:1531
347. Mellor H, Parker PJ (1998) The extended protein kinase C superfamily. Biochem J 332:281
348. Hatton JP, Gaubert F, Cazenave JP, Schmitt D (2002) Microgravity modifies protein kinase C isoform translocation in the human monocytic cell line U937 and human peripheral blood T-cells. J Cell Biochem 87:39
349. Phetsouphanh C, Kelleher AD (2015) The role of PKC-θ in CD4+ T cells and HIV infection: to the nucleus and back again. Front Immunol 6:391
350. Leong I (2018) Role of PKCε in insulin resistance. Nat Rev Endocrinol 14:627
351. Callender JA, Newton AC (2017) Conventional protein kinase C in the brain: 40 years later. Signal 1:NS20160005
352. Kedei N, Lundberg DJ, Toth A, Welburn P, Garfield SH, Blumberg PM (2004) Characterization of the interaction of ingenol 3-angelate with protein kinase C. Cancer Res 64:3243
353. Zhang GG, Kazanietz MG, Blumberg PM, Hurley JH (1995) Crystal structure of the Cys2 activator-binding domain of protein-kinase C-delta in complex with phorbol ester. Cell 81:917
354. Rando RR, Kishi Y (1992) Structural basis of protein kinase C activation by diacylglycerols and tumor promoters. Biochemistry 31:2211
355. Krauter G, Von der Lieth CW, Schmidt R, Hecker E (1996) Structure/activity relationships of polyfunctional diterpenes of the tigliane type. A pharmacophore model for protein-kinase-C activators based on structure/activity studies and molecular modeling of the tumor promoters 12-*O*-tetradecanoylphorbol 13-acetate and 3-*O*-tetradecanoylingenol. Eur J Biochem 242:417
356. Wender PA, Koehler KF, Sharkey NA, Dell'Aquila ML, Blumberg P (1986) Analysis of the phorbol ester pharmacophore on protein kinase C as a guide to the rational design of new classes of analogs. Proc Natl Acad Sci USA 83:4214
357. Katti SS, Krieger IV, Ann J, Lee J, Sacchettini JC, Igumenova TI (2022) Structural anatomy of protein kinase C C1 domain interactions with diacylglycerol and other agonists. Nature Commun 13:2695

358. McConnell DB (2021) Biotin's lessons in drug design. J Med Chem 64:16319
359. Tanaka M, Irie K, Nakagawa Y, Nakamura Y, Ohigashi H, Wender PA (2001) The C-4 hydroxy group of phorbol esters is not necessary for protein kinase C binding. Bioorg Med Chem Lett 11:719
360. Cegelski L (2013) REDOR NMR for drug discovery. Bioorg Med Chem Lett 23:5767
361. Yang H, Staveness D, Ryckbosch SM, Axtman AD, Loy BA, Barnes AB, Pande VS, Schaefer J, Wender PA, Cegelski L (2018) REDOR NMR reveals multiple conformers for a protein kinase C ligand in a membrane environment. ACS Cent Sci 4:89
362. Kwangho N, Wolf-Watz M (2023) Protein dynamics: the future is bright and complicated! Struct Dyn 10:014301
363. Garg R, Benedetti LG, Abera MB, Wang HB, Abba M, Kazaniets MG (2014) Protein kinase C and cancer: what we know and what we do not. Oncogene 33:5225
364. Aquino A, Bianchi N, Terrazzan A, Franzese O (2023) Protein kinase C at the crossroad of mutations, cancer, targeted therapy and immune response. Biology 12:1047
365. Mochly-Rosen D, Das K, Grimes KV (2012) Protein kinase C, an elusive therapeutic target? Nature Rev Drug Discov 11:937
366. De Vries DJ, Herald CL, Pettit GR, Blumberg PM (1988) Demonstration of sub-nanomolar affinity of bryostatin 1 for the phorbol ester receptor in rat brain. Biochem Pharmacol 37:4069
367. Marks F, Gschwendt M (1995) Protein kinase C and skin tumor promotion. Mutat Res 333:161
368. Watson JD (2014) Type 2 diabetes as a redox disease. Lancet 383:841
369. Molenaar RJ, van Noorden CJ (2014) Type 2 diabetes and cancer as redox diseases? Lancet 384:85
370. Usui K, Nakashima C, Takahashi S, Okada T, Ishida Y, Nakajima S, Kitoh A, Nomura T, Dainichi T, Honda T, Katsumoto R, Konishi N, Matsushita M, Otsuka A, Kabashima KJ (2023) TRPV1-positive sensory nerves and neuropeptides are involved in epidermal barrier repair after tape stripping in mice. Allergy Clin Immunol S0091–6749(23):01515–01524
371. Maier M, Olthoff S, Hill K, Zosel C, Magauer T, Wein LA, Schaefer M (2022) KS0365, a novel activator of the transient receptor potential vanilloid 3 (TRPV3) channel, accelerates keratinocyte migration. Br J Pharmacol 179:5290
372. Moore C (2022) The role of TRPV4 channels in cutaneous epithelia. Curr Top Membr 89:139
373. Plant TD, Strotmann R (2007) TRPV4. Handb Exp Pharmacol 179:189
374. Everaerts W, Nilius B, Owsianik G (2010) The vanilloid transient receptor potential channel TRPV4: from structure to disease. Prog Biophys Mol Biol 103:2
375. Watanabe H, Vriens J, Prenen J, Droogmans G, Voets T, Nilius B (2003) Anandamide and arachidonic acid use epoxyeicosatrienoic acids to activate TRPV4 channels. Nature 424:434
376. Atobe S (2019) Activation of transient receptor potential vanilloid (TRPV) 4 as a therapeutic strategy in osteoarthritis. Curr Top Med Chem 19:2254
377. White JPM, Cibelli M, Nagy I, Nilius B, McGeown JG (2023) The emerging pro-algesic profile of transient receptor potential vanilloid type 4. Rev Physiol Biochem Pharmacol 186:57
378. Taivanbat B, Yamazaki S, Nasanbat B, Uchiyama A, Amalia SN, Nasan-Ochir M, Inoue Y, Ishikawa M, Kosaka K, Sekiguchi A, Ogino S, Yokoyama Y, Torii R, Hosoi M, Shibasaki K, Motegi SI (2023) Transient receptor potential vanilloid 4 promotes cutaneous wound healing by regulating keratinocytes and fibroblasts migration and collagen production in fibroblasts in a mouse model. J Dermatol Sci 112:54
379. Yang K, Liu S, Yan H, Lu W, Shan X, Chen H, Bao C, Feng H, Liao J, Liang S, Xu L, Tang H, Yuan JX-J, Zhong N, Wang J (2023) SARS-CoV-2 spike protein receptor-binding domain perturbates intracellular calcium homeostasis and impairs pulmonary vascular endothelial cells. Signal Transduct Target Ther 8:276
380. Klausen TK, Pagani A, Minassi A, Ech-Chahad A, Prenen J, Owsianik G, Hoffmann EK, Pedersen SF, Appendino G, Nilius B (2009) Modulation of the transient receptor potential vanilloid channel TRPV4 by 4α-phorbol esters: a structure-activity study. J Med Chem 52:2933
381. Vitale RM, Avonto C, Del Prete D, Moriello AS, Amodeo P, Appendino G, De Petrocellis L (2020) Discovery of a remarkable methyl shift effect in the vanilloid activity of triterpene amides. J Nat Prod 83:3476

382. Everaerts W, Vriens J, Owsianik G, Appendino G, Voets T, De Ridder D, Nilius B (2010) Functional characterization of transient receptor potential channels in mouse urothelial cells. Am J Physiol Renal Physiol 298:F692
383. Vriens J, Appendino G, Nilius B (2009) Pharmacology of vanilloid transient receptor potential cation channels. Mol Pharmacol 75:1262
384. Szallasi A (2023) Resiniferatoxin: nature's precision medicine to silence TRPV1-positive afferents. Int J Mol Sci 24:15042
385. Appendino G, Cravotto G, Palmisano G, Annunziata R, Szallasi A (1996) Synthesis and evaluation of phorboid 20-homovanillates: discovery of a class of ligands binding to the vanilloid (capsaicin) receptor with different degrees of cooperativity. J Med Chem 39:3123
386. Griffiths CD, Vincent MA, Szallasi A, Colquhoun EQ, Geraghty DP (2000) Functional and desensitizing effects of the novel synthetic vanilloid-like agent 12-phenylacetate 13-acetate 20-homovanillate (PPAHV) in the perfused rat hindlimb. Br J Pharmacol 131:1408
387. Phillips E, Reeve A, Bevan S, McIntyre P (2004) Identification of species-specific determinants of the action of the antagonist capsazepine and the agonist PPAHV on TRPV1. J Biol Chem 279:17165
388. Sharma P, Evans AT, Parker PJ, Evans FJ (1991) NADPH-oxidase activation by protein kinase C-isotypes. Biochem Biophys Res Commun 177:1033
389. Witz G (1991) Active oxygen species as factors in multistage carcinogenesis. Proc Soc Exp Biol Med 198:675
390. Huang R, Zhao L, Chen H, Yin R-H, Li G-Y, Zhan Y-Q, Zhang J-H, Ge C-H, Yu M, Xang X-M (2014) Megakaryocytic differentiation of K562 cells induced by PMA reduced the activity of respiratory chain complex IV. PLoS One 9:e96246
391. Koeffler HP, Bar-Eli M, Territo M (1980) Phorbol diester-induced macrophage differentiation of leukemic blasts from patients with human myelogenous leukemia. J Clin Invest 66:1101
392. Sánchez-Duffhues G, Vo MQ, Pérez M, Calzado MA, Moreno S, Appendino G, Muñoz E (2011) Activation of latent HIV-1 expression by protein kinase C agonists. A novel therapeutic approach to eradicate HIV-1 reservoirs. Curr Drug Targets 12:348
393. Cohen A, Brodie C, Ronit S (2006) An essential role of ERK signalling in TPA-induced reactivation of Kaposi's sarcoma-associated herpesvirus. J Gen Virol 87:795
394. Spivak AM, Planelles V (2018) Novel latency reversing agents for HIV-1 cure. Annu Rev Med 69:421
395. Miana GA, Riaz M, Shahzad-ul-Hussan S, Paracha RZ, Paracha UZ (2015) Prostratin: an overview. Mini Rev Med Chem 15:1122
396. Marquez N, Calzado MA, Sanchez-Duffhues G, Perez M, Minassi A, Pagani A, Appendino G, Diaz L, Muñoz-Fernández MA, Eduardo Muñoz E (2008) Differential effects of phorbol-13-monoesters on human immunodeficiency virus reactivation. Biochem Pharmacol 75:1370
397. Hamer DH, Bocklandt S, McHugh L, Chun TW, Blumberg PM, Sigano DM, Marquez VE (2003) Rational design of drugs that induce human immunodeficiency virus replication. J Virol 77:10227
398. Heywood VH (2011) Ethnopharmacology, food production, nutrition, and biodiversity conservation: towards a sustainable future for indigenous people. J Ethnopharmacol 137:1
399. Cox PA, King S (2024) Bioprospecting. In: Encyclopedia of biodiversity, 3rd edn, vol 3. Elsevier, New York, p 341
400. Cox PA (1997) Nafanua: saving the Samoan rain forest. WH Freeman & Co, New York
401. Johnson HE, Banack SA, Cox PA (2008) Variability in content of the anti-AIDS drug candidate prostratin in Samoan populations of *Homalanthus nutans*. J Nat Prod 71:2041
402. Nothias-Scaglia LF, Pannecouque C, Renucci F, Delang L, Neyts J, Roussi F, Costa J, Leyssen P, Litaudon M, Paolini J (2015) Antiviral activity of diterpene esters on Chikungunya virus and HIV replication. J Nat Prod 78:1277
403. De la Torre-Tarazona HE, Jiménez R, Bueno P, Camarero S, Román L, Fernández-García JL, Beltrán M, Nothias LF, Cachet X, Paolini J, Litaudon M, Alcami J, Bedoya LM (2020) 4-Deoxyphorbol inhibits HIV-1 infection in synergism with antiretroviral drugs and reactivates viral reservoirs through PKC/MEK activation synergizing with vorinostat. Biochem Pharmacol 177:113937

404. Gutiérrez C, Serrano-Villar S, Madrid-Elena N, Pérez-Elías MJ, Martín ME, Barbas C, Ruipérez J, Muñoz E, Muñoz-Fernández MA, Castor T, Moreno S (2016) Bryostatin-1 for latent virus reactivation in HIV-infected patients on antiretroviral therapy. AIDS 30:1385
405. Raghuvanshi R, Bharate SB (2020) Preclinical and clinical studies on bryostatins, a class of marine-derived protein kinase C modulators: a mini-review. Curr Top Med Chem 20:1124
406. Wang M, Sciorillo A, Read S, Divsalar DN, Gyampoh K, Zu G, Yuan Z, Mounzer K, Williams DE, Montaner LJ, de Voogd N, Tietjen I, Andersen RJ, Ansellone J (2022) A potent in vitro and ex vivo HIV-1 latency reversal agent isolated from a *Phorbas* sp. marine sponge. J Nat Prod 85:1274
407. Curreli F, Ahmed S, Benedict Victor SM, Debnath AK (2020) Identification of combinations of protein kinase C activators and histone deacetylase inhibitors that potently reactivate latent HIV. Viruses 12:609
408. Wirasisya DG, Hohmann J (2023) An overview of the traditional use, phytochemistry, and biological activity of the genus *Homalanthus*. Fitoterapia 166:105466
409. Tsai JY, Rédei D, Forgo P, Li Y, Vasas A, Hohmann J, Wu CC (2016) Isolation of phorbol esters from *Euphorbia grandicornis* and evaluation of protein kinase C- and human platelet-activating effects of Euphorbiaceae diterpenes. J Nat Prod 79:2658
410. Gomes TG, Hadi AIIA, Costa Alves GS, Mendonça S, De Siqueira FG, Miller RNG (2018) Currents strategies for the detoxification of *Jatropha curcas* seed cake: a review. J Agric Food Chem 66:2510
411. Nakao M, Hasegawa G, Yasuhara T, Ishihara Y (2015) Degradation of *Jatropha curcas* phorbol esters derived from *Jatropha* oil cake and their tumor-promoting activity. Ecotoxicol Environ Saf 114:357
412. Yamagiwa K, Ichikawa K (1915) Experimentalle Studie über die Pathogenese der Epithelialgeschwülste. Mitt Med Fak Kaiserl Univ Tokyo 15:296
413. Fujiki H (2014) Gist of Dr. Katsusaburo Yamagiwa's papers entitled "Experimental study on the pathogenesis of epithelial tumors" (I to VI reports). Cancer Sci 105:143
414. Garfield S (2001) Mauve: how one man invented a color that changed the world. Faber & Faber, London
415. Cook JW, Hewett CL, Hieger I (1933) The isolation of a cancer-producing hydrocarbon from coal tar. Part I. Concentration of the active substance. J Chem Soc 1:395
416. Slaga TJ (1983) Overview of tumor promotion in animals. Environ Health Perspec 50:3
417. Mottram JC (1944) A developing factor in experimental blastogenesis. J Pathol Bacteriol 56:181
418. Berenblum I (1977) Cancer research in historical perspective: an autobiographical essay. Cancer Res 37:1
419. Fürstenberger G, Berry DL, Sorg B, Marks F (1981) Skin tumor promotion by phorbol esters is a two-stage process. Proc Natl Acad Sci USA 12:7722
420. Ingram AJ, Grasso P (1991) Evidence for and possible mechanisms of non-genotoxic carcinogenesis in mouse skin. Mutat Res 248:333
421. Kinghorn AD, Evans FJ (1975) A biological screen of selected species of the genus *Euphorbia* for skin irritant effects. Planta Med 28:325
422. Dhalendra G, Satapathy T, Roy A (2013) Animal models for inflammation: a review. Asian J Pharm Res 3:207
423. Tubaro A, Dri P, Delbello G, Zilli C, Della Loggia R (1986) The croton oil ear test revisited. Agents Actions 17:347
424. Colorado A, Slama JT, Stavinoha WB (1991) A new method for measuring auricular inflammation in the mouse. J Pharmacol Methods 26:73
425. Kinghorn AD, Harjes KK, Doorenbos NJ (1977) Screening procedure for phorbol esters using brine shrimp (*Artemia salina*) larvae. J Pharm Sci 66:1362
426. Ruebhart DR, Cock IE, Shaw GR (2008) Brine shrimp bioassay: importance of correct taxonomic identification of *Artemia* (Anostraca) species. Environ Toxicol 23:555
427. Roeser H, Doege T, Hecker E (1991) Toxicokinetics of tumour promoters of mouse skin. II. Metabolism of the tumour promoter 12-*O*-tetradecanoylphorbol-13-acetate in mouse skin and biological activities of metabolites. Carcinogenesis 12:1563

428. O'Brien TG, Saladik D (1980) Differences in the metabolism of 12-*O*-[^{3}H]tetradecanoylphorbol-13-acetate and [^{3}H]phorbol-12,13-didecanoate by cells in culture. Cancer Res 40:4433
429. Müller G, Hergenhahn M, Roeser H, Tremp GL, Schmidt R, Hecker E (1991) Toxicokinetics of tumor promoters of mouse skin. I. Metabolism of phorbol and ingenol esters by mouse liver microsomes. Carcinogenesis 11:1127
430. Han ZT, Zhu XX, Yan RYg, Sun JZ, Tian GF, Liu XJ, Cao GS, Newmark HL, Conney AH, Chang RL (1998) Effect of intravenous infusions of 12-*O*-tetradecanoylphorbol-13-acetate (TPA) in patients with myelocytic leukemia: preliminary studies on therapeutic efficacy and toxicity. Proc Natl Acad Sci USA 95:5357
431. Makkar HPS, Becker H (2010) Are *Jatropha curcas* phorbol esters degraded by rumen microbes? J Sci Food Agric 90:1562
432. De Ridder T, Ruppin M, Wheeless M, Williams S, Reddell P (2020) Use of the intratumoural anticancer drug tigilanol tiglate in two horses. Front Vet Sci 7:639
433. Hansen N, Nicholas N, Pack G, Mackie JT, Shipstone M, Munday JS, Reddell P, Orbell G, Malik R (2017) Progressive cutaneous viral pigmented plaques in three Hungarian Vizslas and the response of lesions to topical tigilanol tiglate gel. Vet Med Sci 4:53
434. Miller J, Campbell J, Blum A, Reddell P, Gordon V, Schmidt P, Lowden S (2019) Dose characterization of the investigational anticancer drug tigilanol tiglate (EBC-46) in the local treatment of canine mast cell tumors. Front Vet Sci 6:106
435. De Ridder TR, Campbell JE, Burke-Schwarz C, Clegg D, Elliot EL, Geller S, Kozak W, Pittenger ST, Pruitt JB, Riehl J, White J, Wiest ML, Johannes CM, Morton J, Jones PD, Schmidt PF, Gordon V, Reddell P (2021) Randomized controlled clinical study evaluating the efficacy and safety of intratumoral treatment of canine mast cell tumors with tigilanol tiglate (EBC-46). J Vet Intern Med 35:415
436. Jones PD, Campbell JE, Brown G, Johannes CM, Reddell P (2021) Recurrence-free interval 12 months after local treatment of mast cell tumors in dogs using intratumoral injection of tigilanol tiglate. J Vet Intern Med 35:451
437. Brown GK, Finlay JR, Straw RC, Ziea JY, Leung B, O'Connell K, Thomson MJ, Campbell JE, Jones PD, Reddell P (2022) Treatment of multiple synchronous canine mast cell tumours using intratumoural tigilanol tiglate. Front Vet Sci 9:1003165
438. De Ridder T, Reddell P, Jones P, Brown G, Campbell J (2021) Tigilanol tiglate-mediated margins: a comparison with surgical margins in successful treatment of canine mast cell tumours. Front Vet Sci 8:764800
439. Brown GK, Campbell JE, Jones PD, De Ridder TR, Reddell P, Chad M Johannes CM (2021) Intratumoural treatment of 18 cytologically diagnosed canine high-grade mast cell tumours with tigilanol tiglate. Front Vet Sci 8:675804
440. Wei JJ, Kim HS, Spencer CA, Brennan-Crispi D, Zheng Y, Johnson NM, Rosenbach M, Miller C, Leung DH, Cotsarelis G, Leung TH (2020) Activation of TRPA1 nociceptor promotes systemic adult mammalian skin regeneration. Sci Immunol 5:eaba5683
441. Powell LC, Cullen JK, Boyle GM, De Ridder T, Yap PY, Xue W, Pierce CJ, Pritchard MF, Menzies GE, Abdulkarim M, Adams JYM, Stokniene J, Francis LW, Gumbleton M, Johns J, Hill KE, Jones AV, Parsons PG, Reddell P, Thomas DW (2022) Topical, immunomodulatory epoxy-tiglianes induce biofilm disruption and healing in acute and chronic skin wounds. Sci Transl Med 14:eabn3758
442. Bio-Europe. https://informaconnect.com/bioeurope/sponsors/qbiotics/. Accessed on 14 Jan 2023
443. Save the rhino. https://www.savetherhino.org/rhino-species/black-rhino/tracking-the-desert-adapted-black-rhino-of-namibia/. Accessed on 14 Apr 2024
444. De Beer T, Otto M, Pretoruis B, Schönfeldt HC (2021) Monitoring the quality of honey: South African case study. Food Chem 343:128527
445. Knipper M, Breer H (1987) Protein kinase C in the nervous system of insects: effects of phorbol esters on cholinergic synapses. Neurochem Int 10:323

446. Feyereisen R, Farnsworth DE (1987) Inhibition of insect juvenile hormone synthesis by phorbol 12-myristate 13-acetate. FEBS Lett 222:345
447. Wink M, Grimm C, Koschmieder C, Sporer F, Bergeot O (2000) Sequestration of phorbolesters by the aposematically coloured bug *Pachycoris klugii* (Heteroptera: Scutelleridae) feeding on *Jatropha curcas* (Euphorbiaceae). Chemoecology 10:179
448. Barth MB, Buchwalder K, Kawahara AY, Zhou X, Liu S, Krezdorn N, Rotter B, Horres R, Hundsdoerfer AK (2018) Functional characterization of the *Hyles euphorbiae* hawkmoth transcriptome reveals strong expression of phorbol ester detoxification and seasonal cold hardiness genes. Front Zool 15:20
449. Fiedler S, Langner M, Oertel R, Daniel G, Vassão DG, Nitschke J, Gershenzon J, Hundsdoerfer AK (2023) First insights into herbivore–plant interaction of the spurge hawkmoth and *Euphorbia cyparissias*. Front Ecol Evol 11:1197194
450. Córdova-Albores LC, Bautista-Baños S, Barrera-Necha LL, Salvador-Figueroa M, Adriano-Anaya MDL, Ovando-Medina I (2016) Physic nut (*Jatropha curcas* L.) extracts and active compounds with biological activity against pest insects and pathogenic fungi. In: Betancur-Ancona D, Segura-Campos M (eds) The promising future of *Jatropha curcas*. Properties and potential applications. Nova Science Publishers, Hauppauge, NY, p 87
451. Ratnadass A, Wink M (2012) The phorbol ester fraction from *Jatropha curcas* seed oil: potential and limits for crop protection against insect pests. Int J Mol Sci 13:16157
452. Ito Y, Ohigashi H, Kashimizu K, Yi Z (1983) Epstein-Barr virus activating principle in the ether extracts of soils collected under plants which contain active diterpenoid esters. Cancer Lett 19:113
453. Francis G, Makkar HPS, Carle R, Mittelbach M, Wink M, Herrera JM, Kodekalra R, Becker K (2021) Critique on conclusions regarding toxic compounds in *Jatropha curcas* kernel cake. Nat Commun Biol 4:1348
454. Gordon RJ, Hungerford NL, Laycock B, Fletcher MT (2020) A review of *Pimelea* poisoning of livestock. Toxicon 186:46
455. Becker K, Makkar HP (1998) Effects of phorbol esters in carp (*Cyprinus carpio* L). Vet Hum Toxicol 40:82
456. Hetter GP (2000) An examination of the phenol-croton oil peel: part II. The lay peelers and their croton oil formulas. Plast Reconstr Surg 105:240
457. Scuderi N, Toth BA (eds) (2016) International textbook of aesthetic surgery. Springer, Berlin, Heidelberg, p 841
458. Wambier CG, MD, Lee KC, Soon SL, Sterling JB, Rullan PP, MD, Landau M, Brody HJ (2019) Advanced chemical peels: phenol-croton oil peel. J Am Acad Dermatol 81:327
459. Hetter GP (2000) An examination of the phenol-croton oil peel: part I. Dissecting the formula. Plast Reconstr Surg 105:227
460. De Morais e Silva Cardoso FA, de Deus Moura R, Ferreira Salles Pilar E, Gomes Moura IC, Amarante Miot H, da Costa A (2022) Phenol–croton oil peel enhances type-1 and type-3 collagen amounts by stimulating SIRT-6 and SIRT-7. Int J Dermatol 61:e71
461. Hetter GP (2000) An examination of the phenol-croton oil peel: part III. The plastic surgeons' role. Plast Reconstr Surg 105:75
462. Hetter GP (2000) An examination of the phenol-croton oil peel: part IV. Face peel results with different concentrations of phenol and croton oil. Plast Reconstr Surg 105:1061
463. Eberbach plastic surgery. https://eberbach.com/plastic-surgery-procedures-tampa-bay-florida/face/croton-oil-peel. Accessed 24 May 2024
464. Baker TJ (2002) Is the phenol-croton oil peel safe? Plast Reconstr Surg 110:715
465. NIH National Center for Advancing Translational Sciences. https://drugs.ncats.io/drug/NI40JAQ945. Accessed 15 Apr 2024
466. Schaar D, Goodell L, Aisner J, Cui XX, Han ZT, Chang R, Martin J, Grospe S, Dudek L, Riley J, Manago J, Lin Y, Rubin EH, Conney A, Strair RK (2006) A phase I clinical trial of 12- *O*-tetradecanoylphorbol-13-acetate for patients with relapsed/refractory malignancies. Cancer Chemother Pharmacol 57:789

467. Fang B, Song Y, Han Z, Wei X, Lin Q, Zhu X, Yang R, Sun J, Tian G, Liu X, Cao G, Shi Y, Nie N, Li D, Zhao RC (2007) Synergistic interactions between 12-*O*-tetradecanoylphorbol-13-acetate (TPA) and imatinib in patients with chronic myeloid leukemia in blastic phase that is resistant to standard-dose imatinib. Leuk Res 31:1441
468. Pegoraro L, Abraham J, Cooper RA, Levis A, Lange B, Meo P, Rovera G (1980) Differentiation of human leukemias in response to 12-*O*-tetradecanoylphorbol-13-acetate in vitro. Blood 55:859
469. National Library of Medicine. https://clinicaltrials.gov/study/NCT01009931. Accessed 10 Jan 2024
470. Panizza BJ, de Souza P, Cooper A, Roohullah A, Karapetis CS, Lickliter JD (2019) Phase I dose-escalation study to determine the safety, tolerability, preliminary efficacy and pharmacokinetics of an intratumoral injection of tigilanol tiglate (EBC-46). EBioMedicine 50:433
471. Turner R (1995) Euphorbias. A gardeners' guide. Batsford Ltd, London
472. Zayed SM, Farghaly M, Taha H, Gotta H, E Hecker E (1988) Dietary cancer risk conditional cancerogens in produce of livestock fed on species of spurge (Euphorbiaceae). I. Skin irritant and tumor-promoting ingenane-type diterpene esters in *E. peplus*, one of several herbaceous *Euphorbia* species contaminating fodder of livestock. J Cancer Res Clin Oncol 124:131
473. Hecker E (1984) Cocarcinogens of the tumour-promoter type as potential risk factors of cancer in man. A first complete experimental analysis of an etiological model situation and some of its consequences. IARC Sci Publ 56:441
474. Bracher F, Eisenreich WJ, Mühlbacher J, Dreyer M, Bringmann G (2004) Saludimerines A and B, novel-type dimeric alkaloids with stereogenic centers and configurationally semistable biaryl axes. J Org Chem 69:8602
475. Morton JF (1971) Welensali (*Croton flavens*): folk uses and properties. Econ Bot 25:457
476. European Food Safety Agency (EFSA) Compendium of Botanicals. https://www.efsa.europa.eu/en/data-report/compendium-botanicalsbotanicals
477. Tóth-Soma LT, Gulyás S, Szegletes Z (1993) Functional connection between intracellular and extracellular secretion in species of *Euphorbia* genus. Acta Biol Hung 44:433
478. Juritz CF (1925) The problem of noors honey. Agric J South Africa 10:334
479. Upadhyay RR, Islampanah S, Davoodi A (1980) Presence of a tumor-promoting factor in honey. Gan 71:557
480. Yan S, Wang K, Al Naggar Y, Vander Heyden Y, Zhao L, Wu L, Xue X (2022) Natural plant toxins in honey: an ignored threat to human health. J Hazard Mater 424:127682
481. Boutoub O, El-Guendouz S, Estevinho LM, Paula VB, Aazza S, El Ghadraoui L, Rodrigues B, Raposo S, Carlier J, Costa MC, Miguel MG (2021) Antioxidant activity and enzyme inhibitory potential of *Euphorbia resinifera* and *E. officinarum* honeys from Morocco and plant aqueous extracts. Environ Sci Pollut Res Int 28:503
482. Appendino G, Szallasi A (1997) *Euphorbium*: modern research on its active principle, resiniferatoxin, revives an ancient medicine. Life Sci 60:681
483. Anonymous (1937) Gesundheitsschädigung durch Holzöl. Pharm Z 82:725
484. Strichland NH (2000) Eating a manchineel "beach apple." Br Med J 321:428
485. Cooper MR, Johnson AW (1984) Poisonous plants in Britain and their effects on animals and man. Reference Book 161, Ministry of Agriculture, Fisheries and Food, London
486. Sofat BK, Sood GC, Chandel RD, Mehrotra SK (1972) *Euphorbia royleana* latex keratitis. Am J Opthalmol 74:634
487. Kinghorn AD (1979) Cocarcinogenic irritant Euphorbiaceae. In: Kinghorn AD (ed) Toxic plants. Columbia University Press, New York, p 137
488. The Guiness Book of World Records. https://www.guinnessworldrecords.com/world-records/most-dangerous-tree. Accessed 10 Jan 2024
489. Adolf W, Hecker E (1984) On the active principles of the spurge family, X. Skin irritants, cocarcinogens, and cryptic cocarcinogens from the latex of the manchineel tree. J Nat Prod 47:482

490. Ernst M, Grace OM, Saslis-Lagoudakis CH, Nilsson N, Simonsen HT, Rønsted N (2015) Global medicinal uses of *Euphorbia* L. (Euphorbiaceae). J Ethnopharmacol 176:90
491. Frohn A, Frohn C, Steul KP, Thie J-H (1993) Wolfsmilchverätzung. Ophthalmologe 90:58
492. Anonymous (2006) *Euphorbia*. Drugs and Lactation Database (LactMed®), National Institute of Child Health and Human Development, Bethesda, MD. https://www.ncbi.nlm.nih.gov/books/NBK501883/. Accessed 10 Jan 2024
493. Using veterinary to human parallel development to speed drugs to market. https://www.nature.com/articles/d43747-022-00076-8. Accessed 10 Jan 2024

Giovanni Appendino obtained a Laurea degree in 1979 from the University of Turin, Italy, where he next held the position of Lecturer and Associated Professor. From 2000 until to his retirement in 2022, he was a Full Professor at the University of Eastern Piedmont in Novara, and is now Professor Emeritus. His research activity takes inspiration from natural products to address issues of relevance to organic, bioorganic, and medicinal chemistry. The Italian Chemical Society awarded him the Quilico Medal in 2009 and the Piria Medal in 2021, and in 2018 he became a member of the Italian Academy of Science (Accademia dei XL). For his contributions to the study of bioactive natural products, in 2023 he was awarded the Phytochemical Society of Europe Silver Medal and the GA Egon Stahl Medal in Gold Award.

Simone Gaeta obtained his M.Sc. and Ph.D. degrees from the University of Eastern Piedmont in 2016 and 2020, respectively, under the supervision of Prof. Giovanni Appendino. After postdoctoral experience at the University of Zürich (UZH) and again at the University of Eastern Piedmont, he joined the Australian life sciences company QBiotics Group Ltd as a Senior Research Chemist. His research activities are focused on the phytochemical identification and characterization of bioactive natural products from the Queensland rainforest and the exploration of chemical space around tigliane diterpenoids.

SPRINGER NATURE

GPSR Compliance

The European Union's (EU) General Product Safety Regulation (GPSR) is a set of rules that requires consumer products to be safe and our obligations to ensure this.

If you have any concerns about our products, you can contact us on ProductSafety@springernature.com

In case Publisher is established outside the EU, the EU authorized representative is:

Springer Nature Customer Service Center GmbH
Europaplatz 3
69115 Heidelberg, Germany

The manufacturer's authorised representative in the EU is Springer Nature Customer Service Centre GmbH, Europaplatz 3, 69115 Heidelberg, Germany. If you have any concerns regarding our products, please contact ProductSafety@springernature.com

Printed and bound by CPI Group (UK) Ltd, Croydon, CR0 4YY
01/10/2026
02225954-0012